CW01670431

Maths — No Problem!

English National Curriculum

Consultant Editors
Dr Wong Khoon Yoong
Dr Julie Alderton

Series Editors
Adam Gifford
Andy Psarianos

Authors
Adam Gifford
Michelle Hayfron

MATHS NO PROBLEM!

Published by Maths — No Problem!

Printed in the United Kingdom
First Printing, 2022

ISBN 978-1-913458-46-1

Maths — No Problem!
hello@mathsnoproblem.com
www.mathsnoproblem.com

Acknowledgements
The publisher would like to thank Dr Anne Hermanson, Alex Laurie and Judy Hornigold for their contributions.

Preface

Welcome to Maths — No Problem!

Maths — No Problem! is a comprehensive series of textbooks and workbooks supported by detailed online teacher guidance to deliver maths mastery in primary schools. The materials use a spiral design with carefully built-up mathematical concepts and processes adapted from the maths mastery approaches used in Singapore. An integral part of the learning process is the Concrete, Pictorial, Abstract (CPA) approach, which forms the backbone of the materials developed for this series.

Maths — No Problem! incorporates the use of concrete aids and manipulatives, problem solving, group work, individual practice and journaling.

Meet our Maths — No Problem! friends who will join you on your journey through this textbook.

Ruby Elliott Amira Charles Lulu Sam Oak Holly Ravi Emma Jacob Hannah

The Maths — No Problem! approach is exemplified throughout the chapters by the following teaching elements:

Chapter Opener

The chapter opener depicts familiar events or contexts that serve as an introduction for pupils to the concept covered in the chapter. It can also be used as an additional prompt for journal writing or an additional exercise for pupils.

Explore

Explore is an introductory activity for pupils, which includes questions related to various lesson objectives.

Master

Each lesson contains a Master section that develops lesson objectives by introducing new concepts through a CPA approach using manipulatives and engaging pictures. Guided examples are provided to reinforce the learning objective.

Activity Time

Activity Time provides pupils with opportunities to work as individuals or in small groups to explore mathematical concepts through games and investigations.

Guided Practice

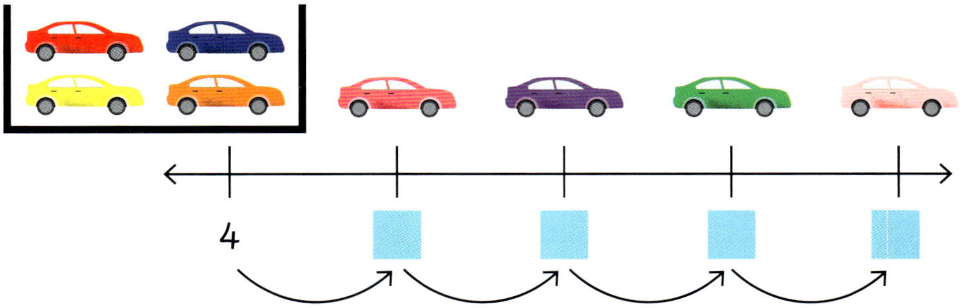

Guided Practice contains questions for further consolidation and the immediate evaluation of pupils' learning.

Mind Challenge

Mind Challenges are non-routine questions for pupils to apply relevant problem-solving skills and develop higher-order thinking skills.

My Maths Journal

My Maths Journal provides pupils with opportunities to show their understanding of the mathematical concepts learnt and to record their learning.

Review

Review allows pupils to reflect on their own learning after each chapter, with support from the teacher.

Review

I can...

- ☐ compare the height and length of objects.
- ☐ measure the height and length of objects.

Contents

Chapter 1	Numbers to 10	Page
1	Counting to 10	2
2	Counting Objects to 10	6
3	Writing to 10	10
4	Counting to Zero	15
5	Comparing Numbers of Objects	18
6	Ordering Numbers	22
7	Comparing Numbers	25
	Chapter Consolidation	28
Chapter 2	**Number Bonds**	
1	Making Number Bonds	30
2	Making Number Stories	34
	Chapter Consolidation	38
Chapter 3	**Addition within 10**	
1	Add by Using Number Bonds	40
2	Add by Counting On	44
3	Completing Number Sentences	48
4	Making Addition Stories	52
5	Solving Picture Problems	55
	Chapter Consolidation	58

Chapter 4 Subtraction within 10

		Page
1	Subtract by Crossing Out	60
2	Subtract by Using Number Bonds	63
3	Subtract by Counting Back	66
4	Making Subtraction Stories	69
5	Solving Picture Problems	73
6	Addition and Subtraction	76
	Chapter Consolidation	79

Chapter 5 Positions

1	Naming Positions	82
2	Naming Positions in Queues	85
3	Naming Left and Right Positions	87
	Chapter Consolidation	91

Chapter 6 Numbers to 20

1	Counting to 20	94
2	Writing to 20	99
3	Comparing Numbers	103
4	Ordering Numbers	107
5	Number Patterns	112
	Chapter Consolidation	116

Chapter 7 Addition and Subtraction within 20

		Page
1	Add by Counting On	118
2	Add by Making 10	121
3	Add by Adding Ones	124
4	Subtract by Counting Back	126
5	Subtract by Subtracting Ones	128
6	Subtract from 10	132
7	Addition and Subtraction Facts	134
	Chapter Consolidation	137

Chapter 8 Shapes and Patterns

1	Recognising 3D Shapes	140
2	Recognising 2D Shapes	144
3	Grouping 2D Shapes	149
4	Making Patterns	154
	Chapter Consolidation	158

Chapter 9 Height and Length

1	Comparing Height and Length	162
2	Measuring Length Using Things	167
3	Measuring Height and Length Using Body Parts	172
4	Measuring Height and Length Using a Ruler	178
	Chapter Consolidation	182

Chapter 1
Numbers to 10

Counting to 10

Lesson 1

Explore

Can you help Emma by placing the cards in order from 0 to 10?

Master

1

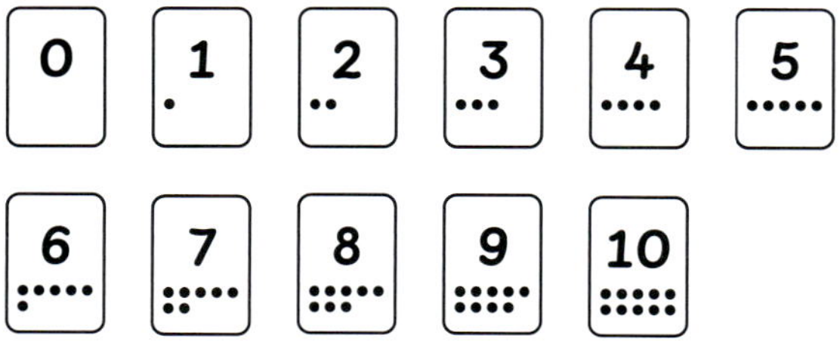

2

3

Start counting from 1 and stop counting at 8.

1, 2, 3, 4, 5, 6, 7, 8

4

Can you count on from 3?

3, 4, 5, 6, 7, 8, 9, 10

5

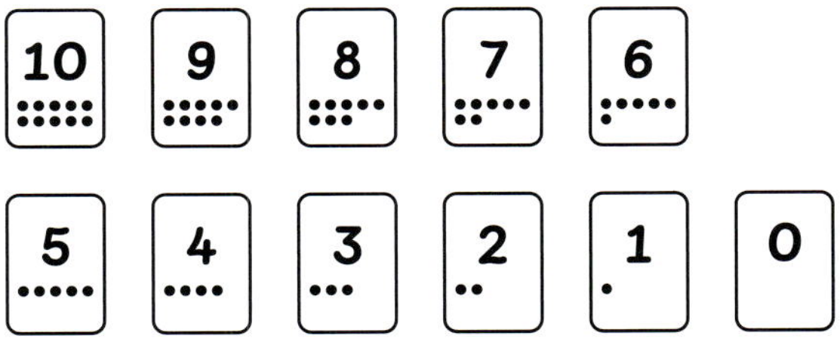
10
9
8
7
6
5
4
3
2
1
0

Let's count back from 10.

10, 9, 8, 7, 6, 5, 4, 3, 2, 1, 0

6

Can you count back from 7?

7, 6, 5, 4, 3, 2, 1, 0

Guided Practice

Help the children by counting.

1

2

3

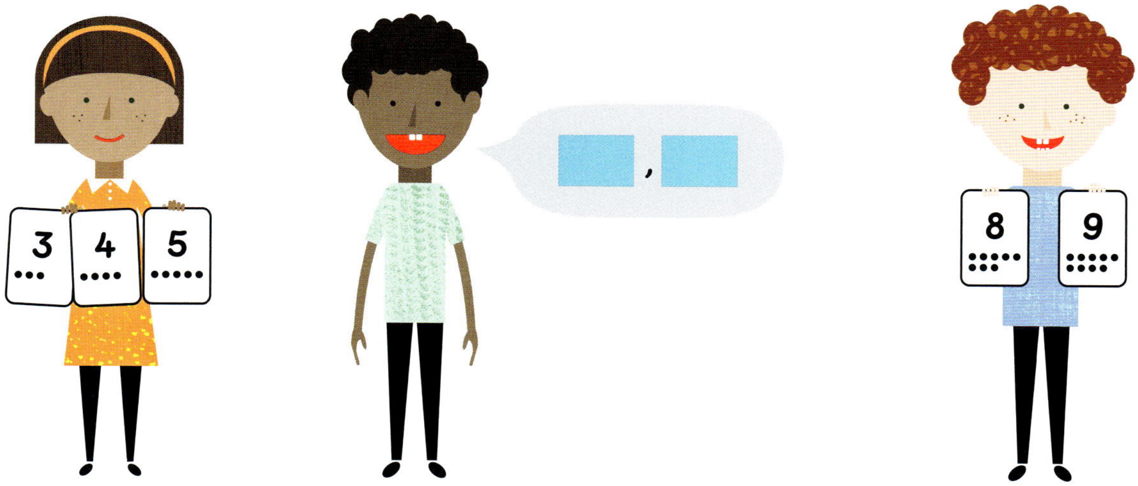

4

5

6

Complete Worksheet **1**: pages **1–2**

Counting Objects to 10

Lesson 2

Explore

Master

1

1

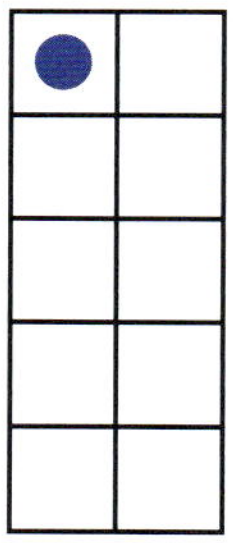

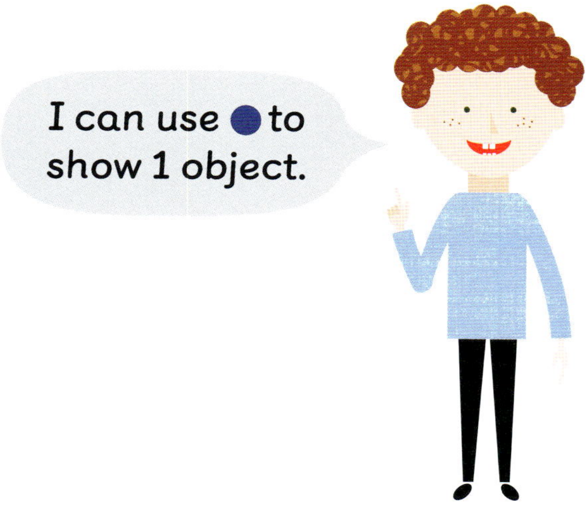

2

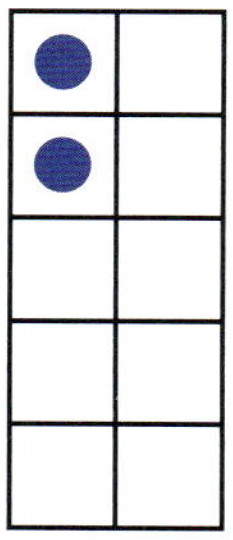

3

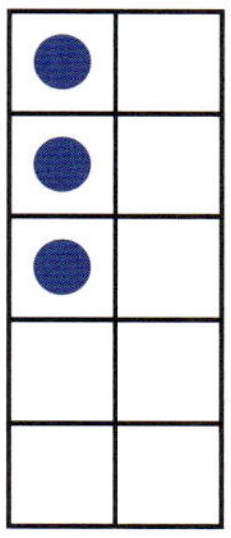

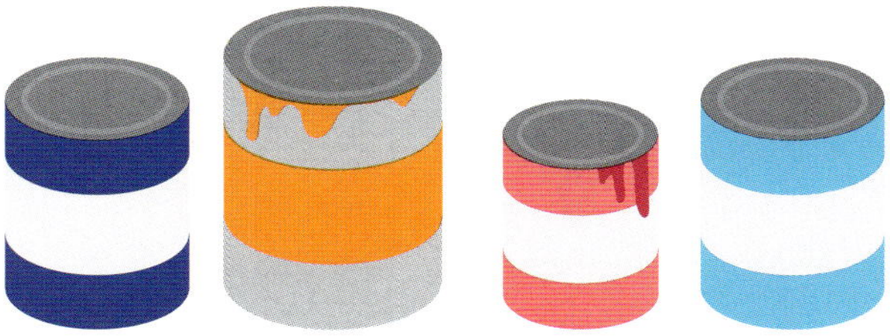

4

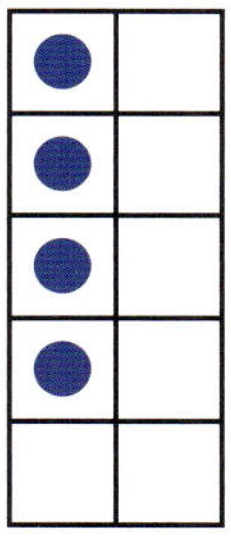

5

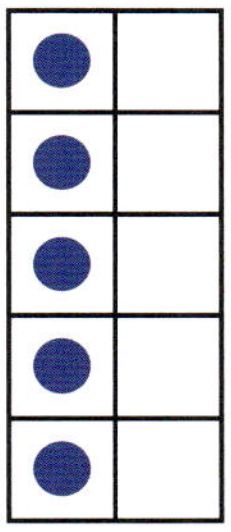

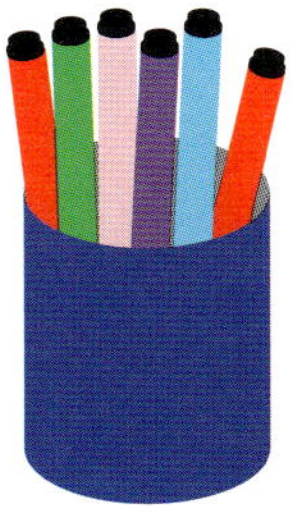

6

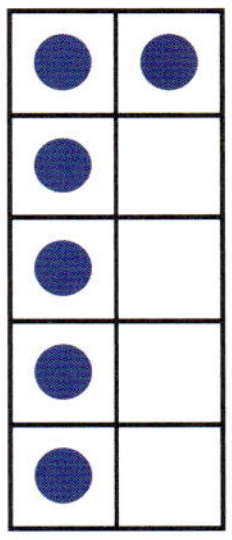

7

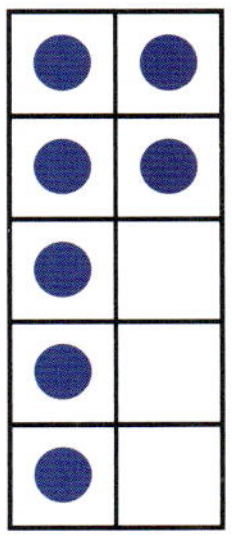

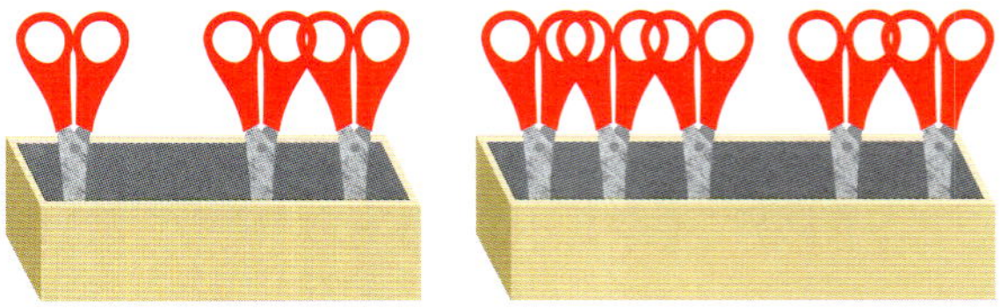

8

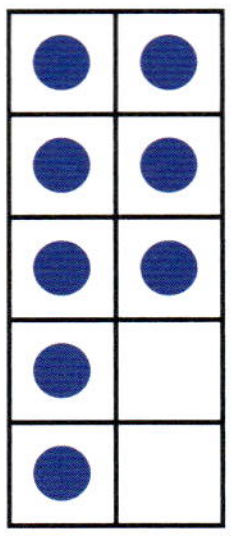

9

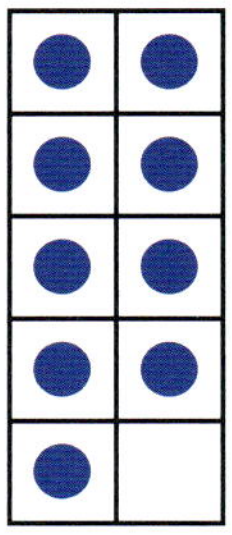

10

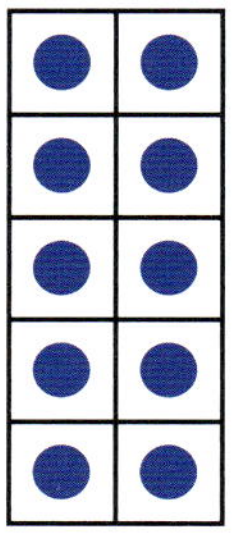

2

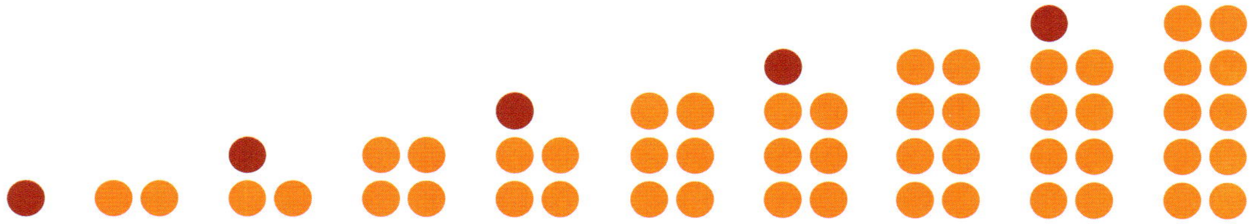

1 2 3 4 5 6 7 8 9 10

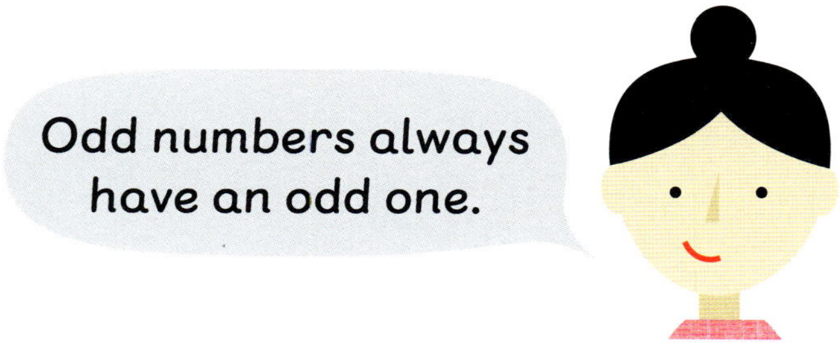

Guided Practice

1 Count.

(a)

(b)

(c)

(d)

2 Match the numbers.

4  • • 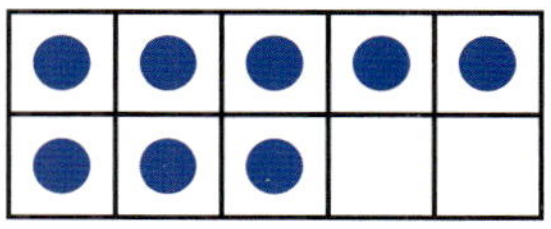

9  • • 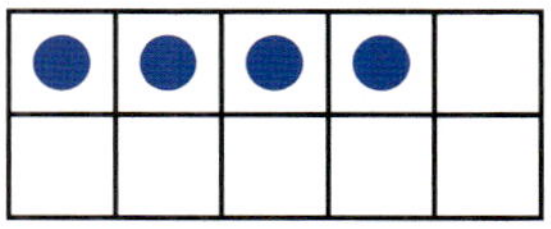

8  • • 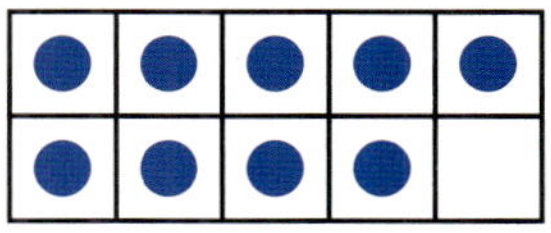

6  • • 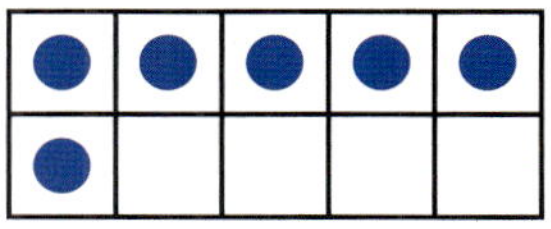

Complete Worksheet **2**: pages **3–6**

Writing to 10

Explore

How many books does Amira have?
Can you help her by counting them?

Master

Help Amira by counting the objects.

1

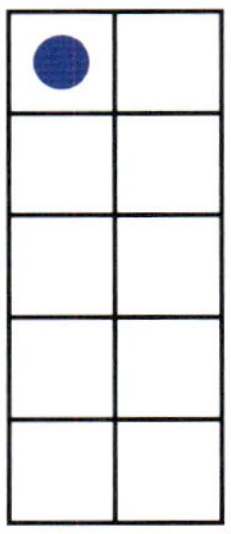

one

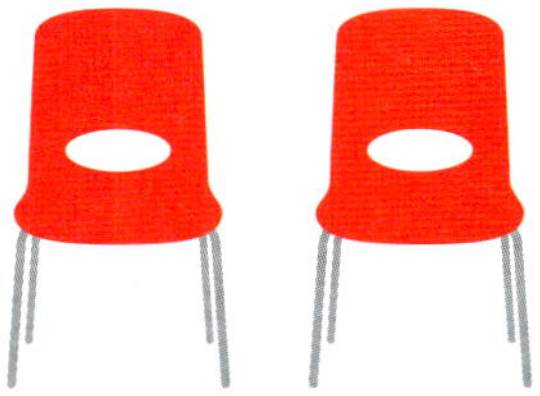

2

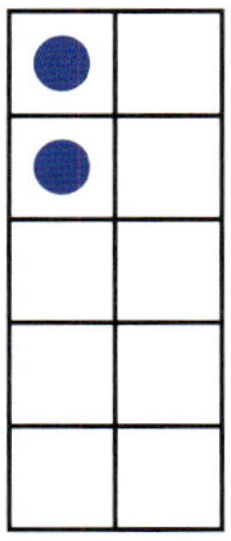

two

3

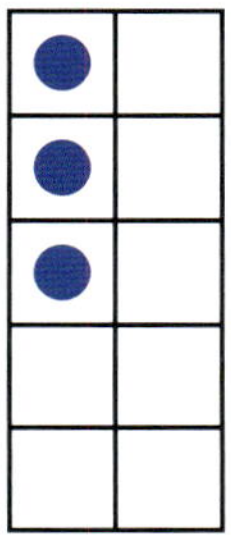

three

4

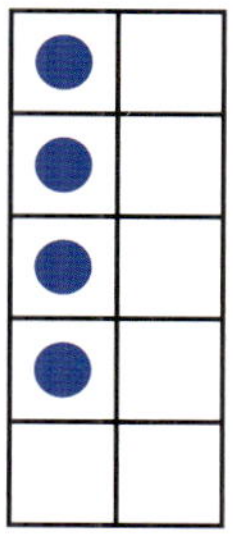

four

5

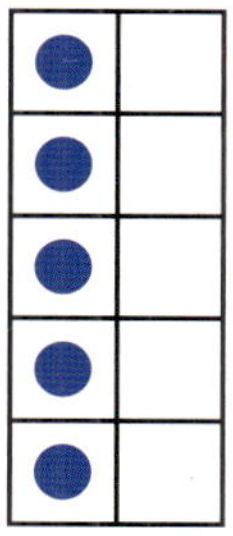

five

6

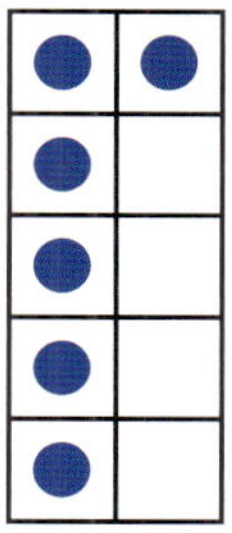

six

7

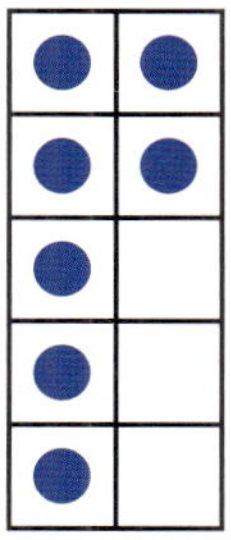

seven

8

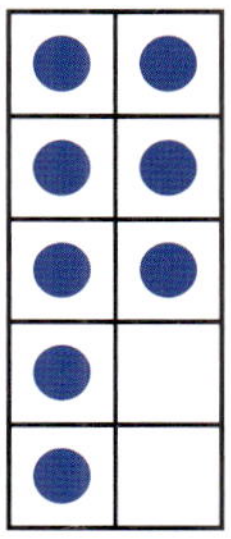

eight

9

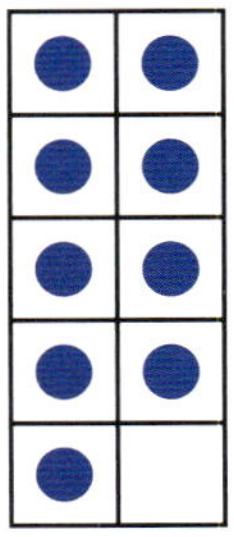

nine

10

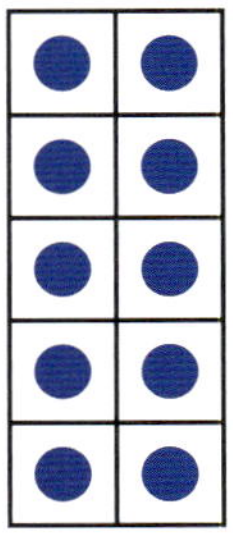

ten

Activity Time

Partner work

You will need:

1. Take a card.
2. Say the number.
3. Find the same number of objects as the number.
4. Leave the card with the objects.
5. Repeat the activity, taking turns.

Guided Practice

1 Help Charles by counting the books.
Write the numbers in words.

(a)

______ books

(b)

______ books

(c)

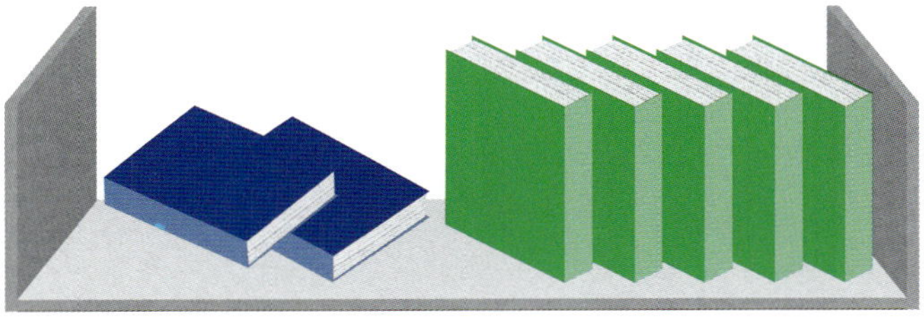

______ books

(d)

______ books

2 Count and write the numbers in words.

(a) three tomatoes

(b) ______ lettuces

(c) ______ carrots

(d) ______ bees

(e) ______ snails

(f) ______ apples

Complete Worksheet **3**: pages **7–10**

Counting to Zero

Lesson 4

Explore

Ravi's friends each borrowed one of his pens.
How many pens are left in the pot?

Master

4

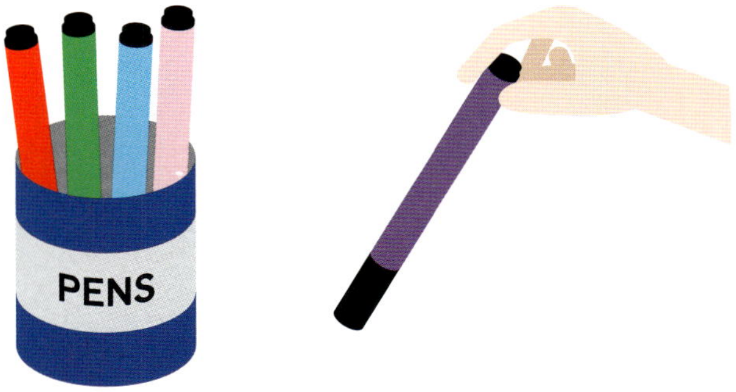

4 pens are left in the pot.

3

3 pens are left in the pot.

2

2 pens are left in the pot.

1

1 pen is left in the pot.

0 

There are no pens left in the pot.
There are zero pens.

Activity Time

Partner work

You will need:

10  

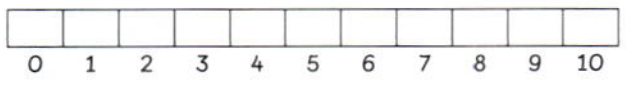

1. Take a .
2. Say the number.
3. Place the same number of cubes on the 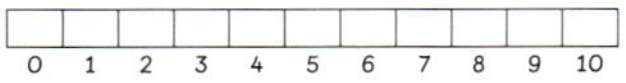.
4. Count on from the number to 10.
5. Ask your partner to count back from the number to 0.
6. Take turns.

Guided Practice

Count back from 10 to 0.

ten	10
nine	9
eight	8
seven	7
six	6
five	5
four	4
three	3
two	2
one	1
zero	0

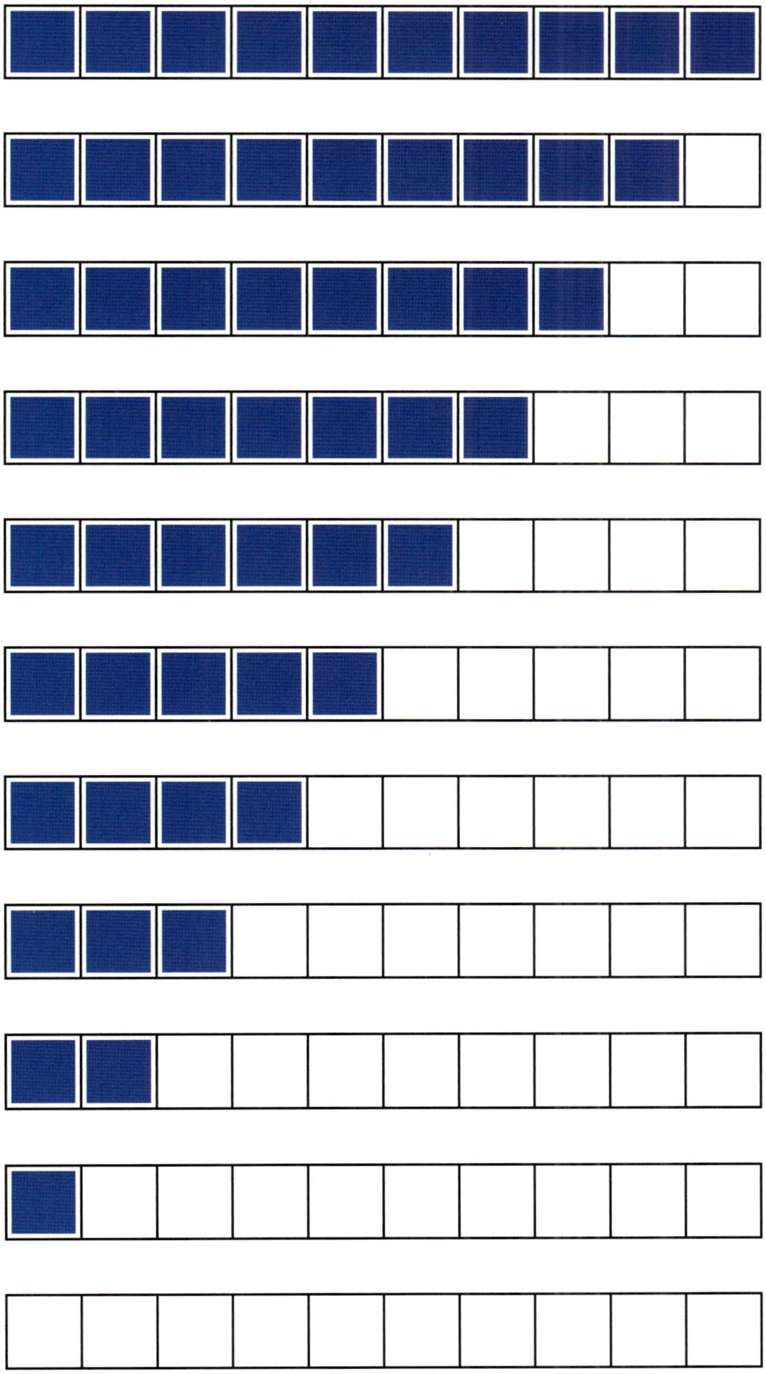

Complete Worksheet 4: pages **11–12**

Comparing Numbers of Objects

Lesson 5

Explore

I think there are more  than .

I think there are more  than .

Who is correct?

Master

How can we check?

1

6 blue tits sit on a branch.
4 finches sit on a branch.
There are fewer finches than blue tits.

2

5 socks are held by 5 pegs on the washing line.
There are as many pegs as socks.
The number of pegs is equal to the number of socks.

3

7 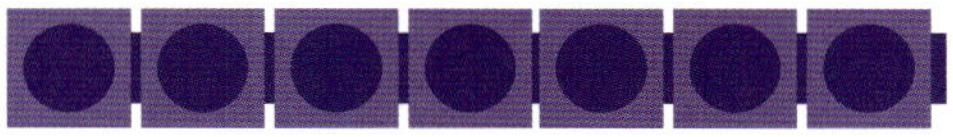

3 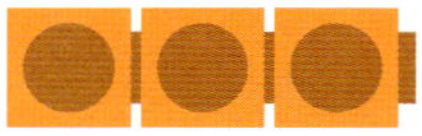

7 is more than 3.

3 is less than 7.

There are more than .

There are fewer than .

Activity Time

Partner work

You will need:
2 2 20

1. Roll 2 .
2. Show each number using and .
3. Tell your partner if a number is more than, less than or equal to the other number.
4. Repeat the activity.

Guided Practice

1 Which row has more animals?

hens

rabbits

2 Which row has fewer sea animals?

crabs

octopuses

3

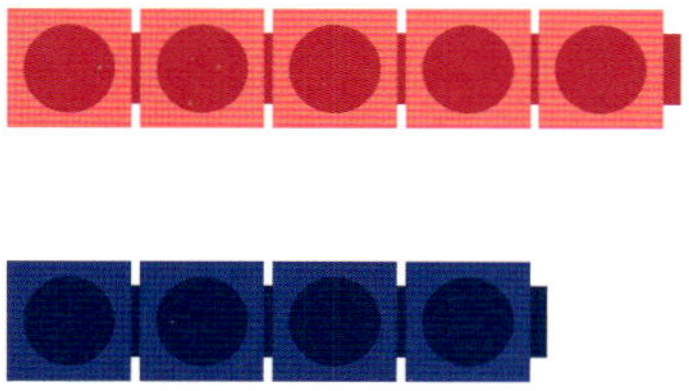

4

5

Complete Worksheet 5: pages 13–14

Ordering Numbers

Lesson 6

Explore

How can we compare the number of items?

Are there more hats or backpacks?
Are there more backpacks or scarves?

Master

1

I can line them up.

2

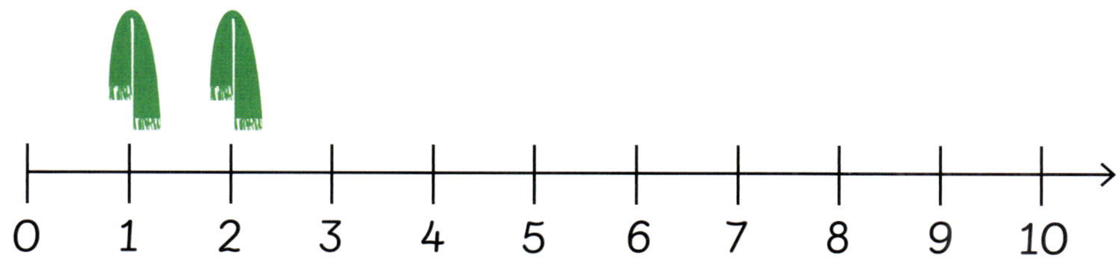

2 scarves

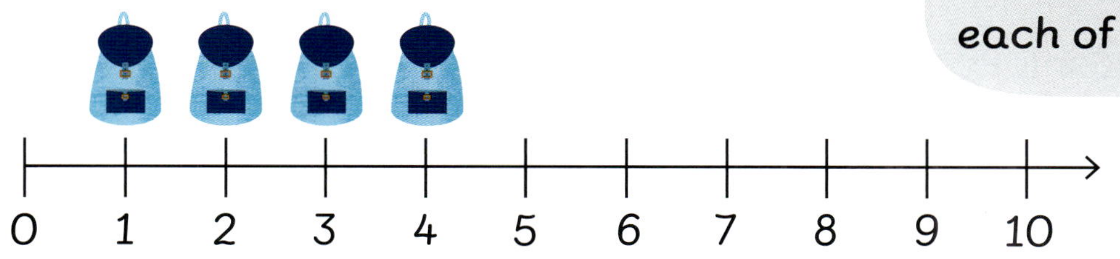

4 backpacks

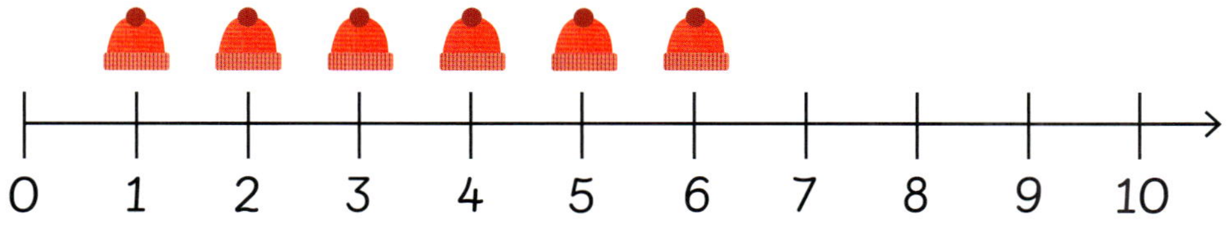

6 hats

6 is more than 2.
6 is more than 4.
6 is the greatest number.

2	4	6

2 is less than 6.
2 is less than 4.
2 is the smallest number.

6	4	2

Guided Practice

1 Which is the smallest number?

10	6	1

______ is the smallest number.

2 Which is the greatest number?

4	9	1

______ is the greatest number.

3 Put the numbers in order.

(a) Begin with the smallest number.

4	6	1

______ ______ ______

(b) Begin with the greatest number.

3	8	5

______ ______ ______

Complete Worksheet **6**: pages **15–16**

Comparing Numbers

Lesson 7

Explore

How many more  does Sam have than Charles?

Master

1  has 7 rubbers.

 has 8 rubbers.

8 is more than 7.

8 is 1 more than 7.

7 is less than 8.

7 is 1 less than 8.

2 Compare 3, 4 and 5.

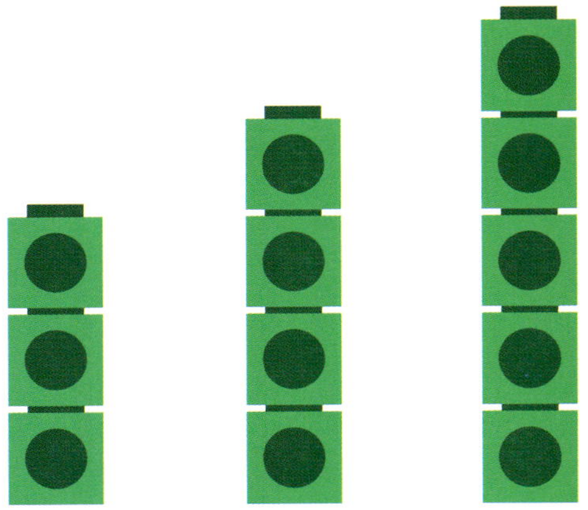

3 4 5

4 is 1 less than 5.
4 is 1 more than 3.
5 is 1 more than 4.
3 is 1 less than 4.

Activity Time

Partner work

You will need:
10 cubes, number cards 1–9

1. Your partner takes a number card and shows it to you.
2. Use cubes to make a number that is 1 more or 1 less.
3. Ask your partner to tell you if it is 1 more or 1 less.
4. Repeat the activity.

Guided Practice

1

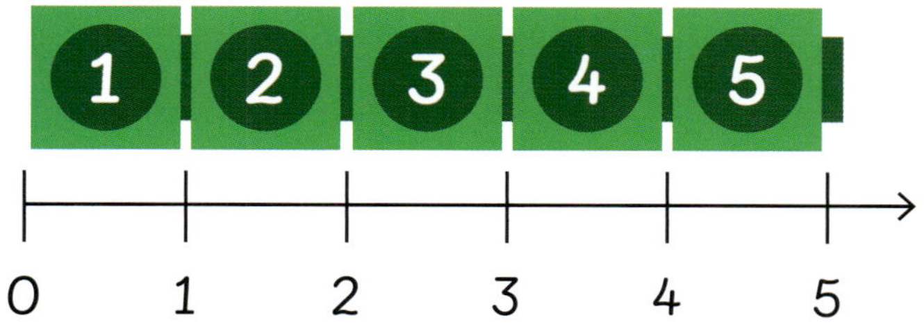

(a) 3 is 1 more than ____.

(b) ____ is 1 more than 4.

(c) ____ is 1 less than 2.

2

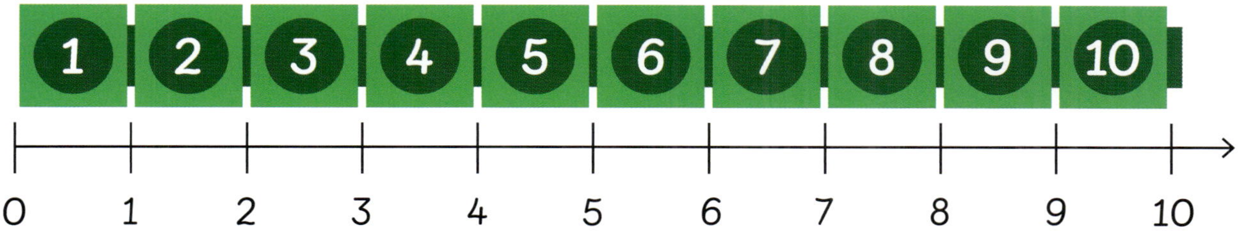

(a) ____ is 1 more than 8.

(b) ____ is 1 less than 10.

(c) ____ is 1 more than 6.

Complete Worksheet **7**: pages **17–18**

Mind Challenge

Each number is attached to a letter.
Write the numbers in order from 1 to 10 and their letters.
The first one has been done for you.

7	2	4	8	5	10	1	6	3	9
↕	↕	↕	↕	↕	↕	↕	↕	↕	↕
A	L	V	T	E	S	I	M	O	H

1	__	__	__	__	__	__	__	__	__
↕									
I	__	__	__	__	__	__	__	__	__

Say the sentence out loud.

My Maths Journal

4	10	6	3	1	7	5	9	2	8

Stick the number cards in your journal in order, starting with 1.
Beside each number, draw the same number of objects and write the number in words.

Review

I can...

- ☐ count from 0 to 10.
- ☐ count backwards from 10 to 0.
- ☐ read and write numbers from 0 to 10.
- ☐ order and compare numbers from 0 to 10.

Can Ravi put the fruit on the plates in different ways?

Chapter 2
Number Bonds

Making Number Bonds

Lesson 1

Explore

How many pieces of fruit are there altogether?
Can we put the fruit on the plates in different ways?

Master

1

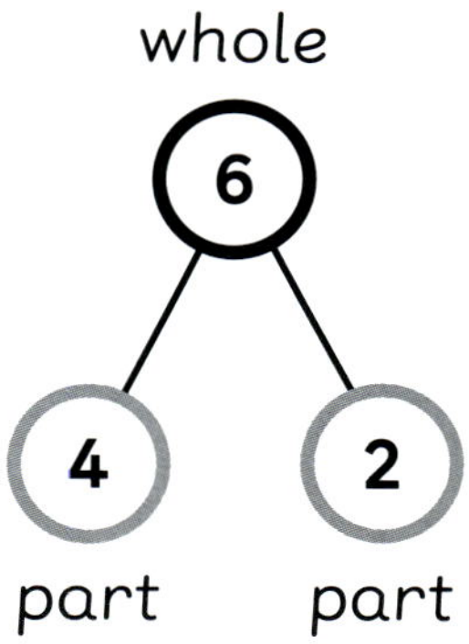

This is called a **number bond**.

2

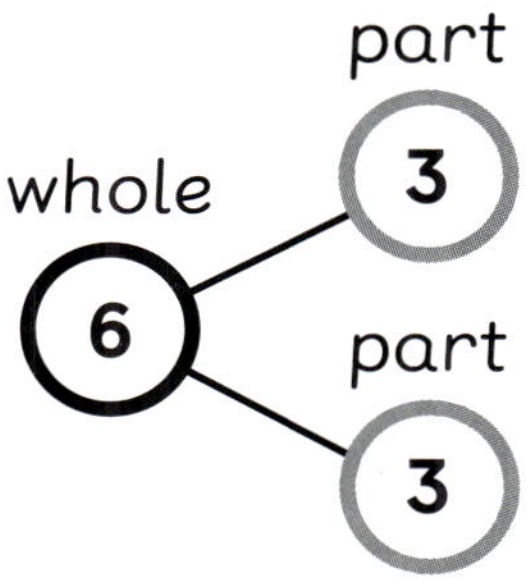

I can see that 6 is made from 3 and 3.

3

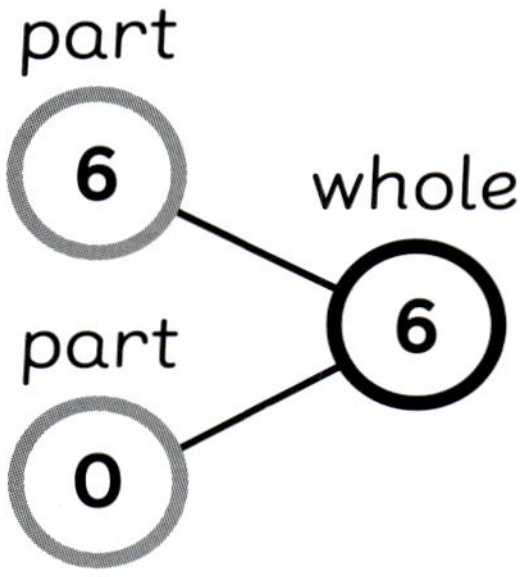

I can see that 6 and 0 make 6.

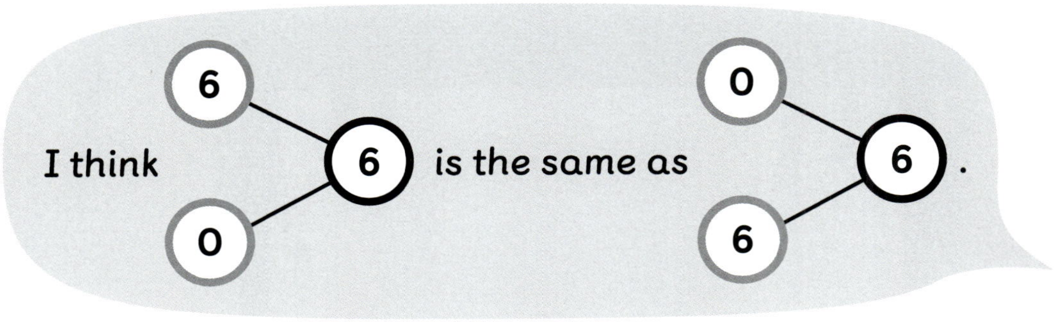

Is Amira correct?

Activity Time

Partner work

You will need:
8

1. Make a number bond using 8 .
2. Ask your partner to write it on a .
3. Repeat the activity, taking turns.

Guided Practice

Complete the number bonds and fill in the blanks.

1

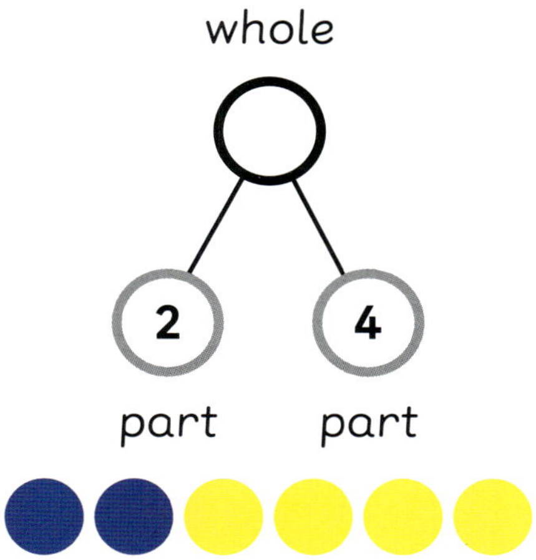

2 and 4 make ___.

2

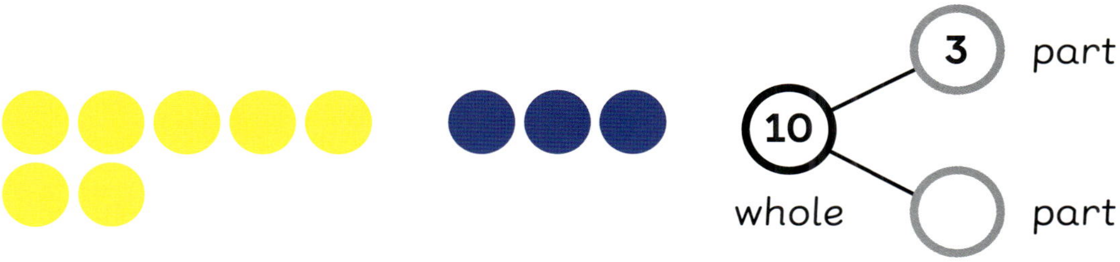

___ and 3 make 10.

3

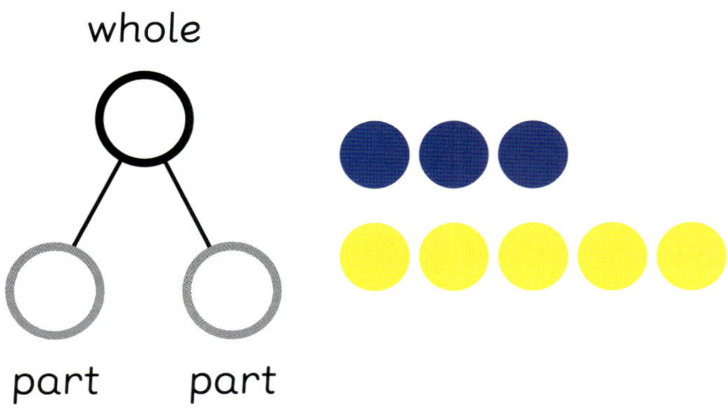

___ and ___ make ___.

Complete Worksheet **1**: pages **25–26**

Making Number Stories

Lesson 2

Explore

Use the picture to tell a number story.

Master

1 This is a number story.

There are 7 animals on the lily pads.

There are 4 frogs on the lily pads.

There are 3 dragonflies on the lily pads.

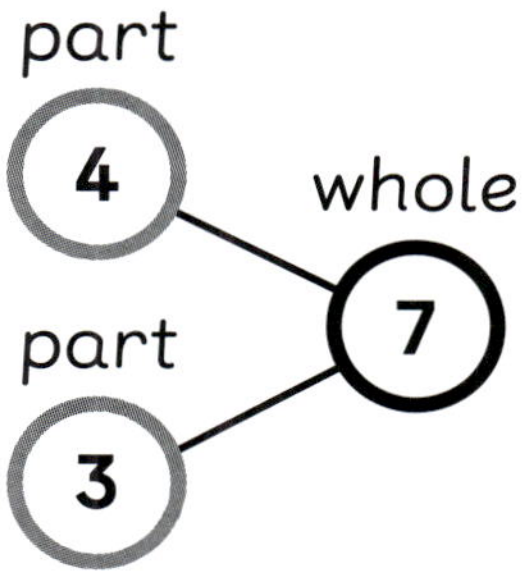

2 There are ____ flowers.

There are ____  .

There are ____  .

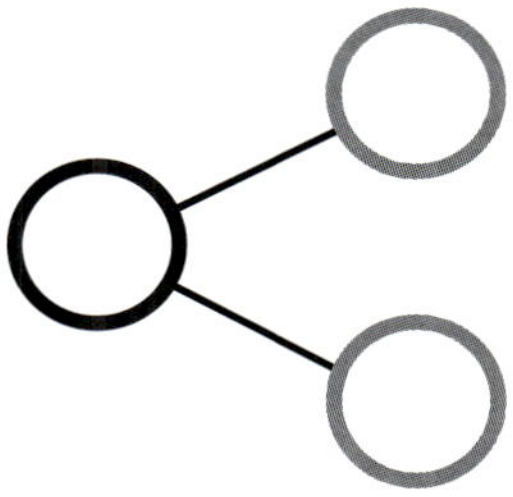

3 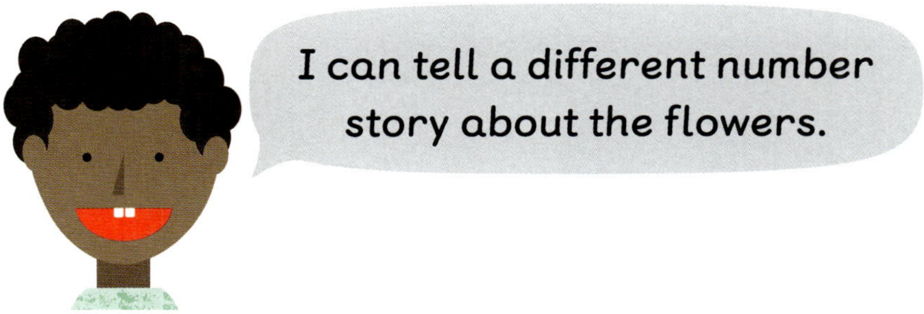

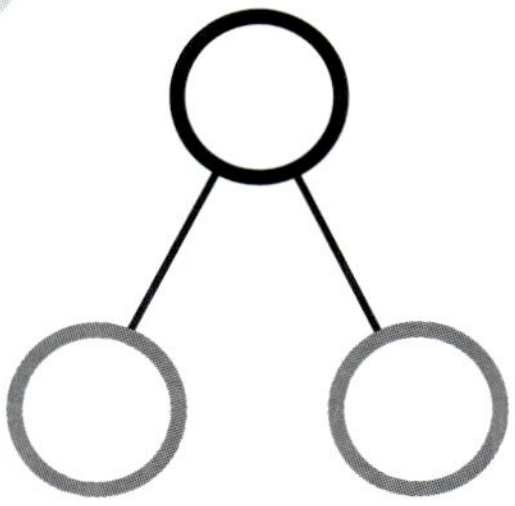

____ flowers are on lily pads.

____ flowers are not on lily pads.

Activity Time

Partner work

You will need:
10

1. Make a number using .
2. Ask your partner to write it on a .
3. Tell a number story using the number bond.
4. Take turns.

Guided Practice

Complete the sentences.

1 There are 7 flowers.

____ are on lily pads.

____ are on the grass.

2 There are ____ birds.

(a) ____ birds are not in the pond.

____ birds are in the pond.

(b) ____ birds are in trees.

____ birds are not in trees.

Complete Worksheet **2**: pages **27–28**

Mind Challenge

Sam started to write a number bond like this and got a bit stuck.
Can you help him by finishing it?

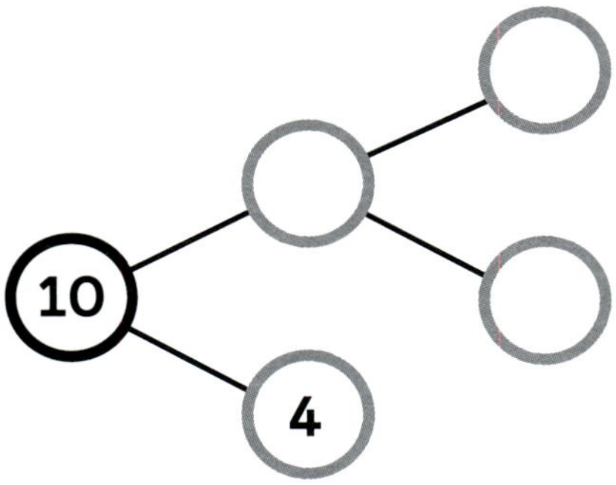

My Maths Journal

Use the frogs in the picture to tell number stories.

I can...

Review

- ☐ make number bonds with numbers up to 10.
- ☐ make number stories with numbers up to 10.

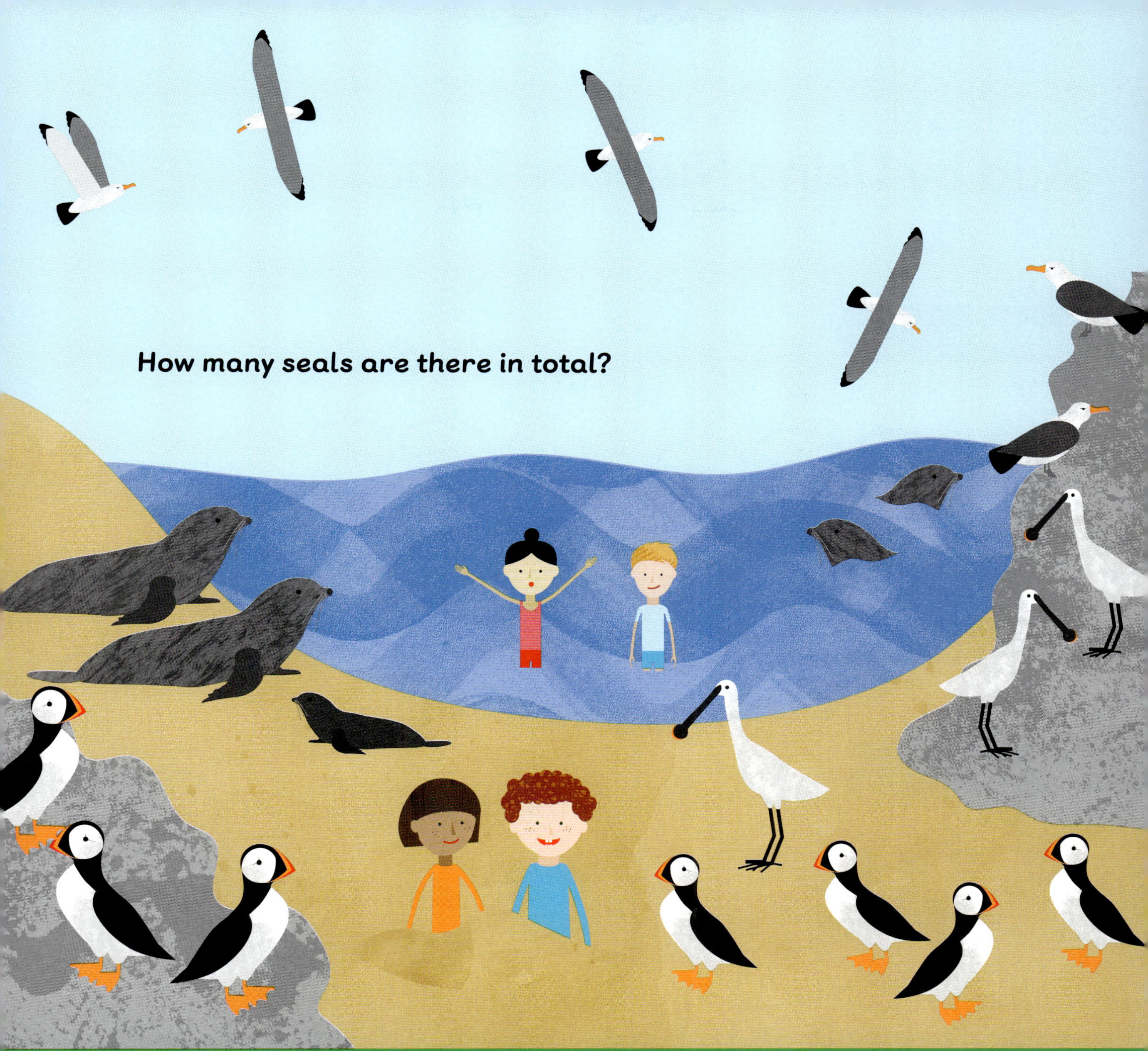

Chapter 3
Addition within 10

Add by Using Number Bonds

Lesson 1

Explore

How many buckets are on the beach?

Master

1

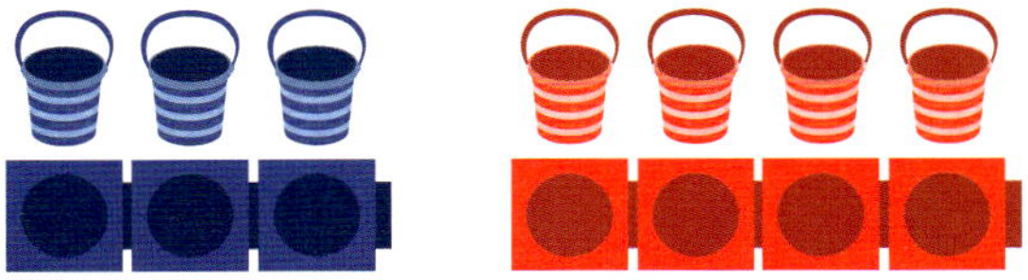

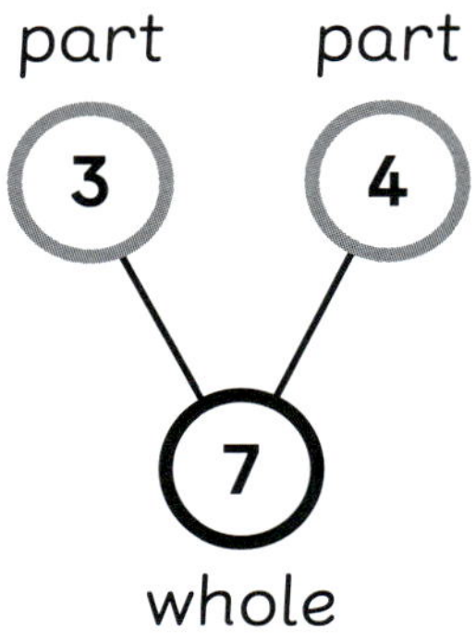

3 + 4 = 7

There are 7 buckets on the beach.

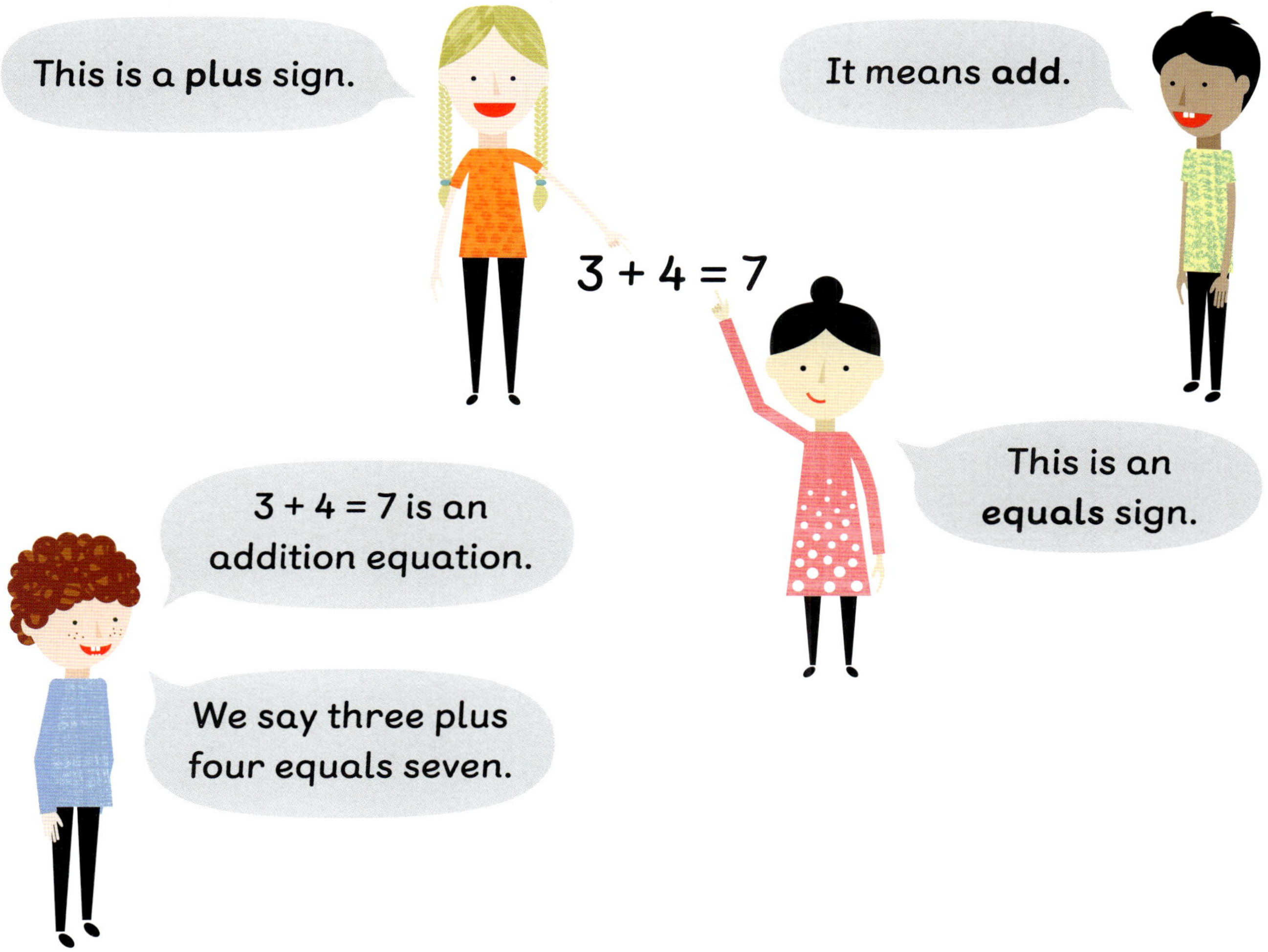

2

6 + 2 = 8

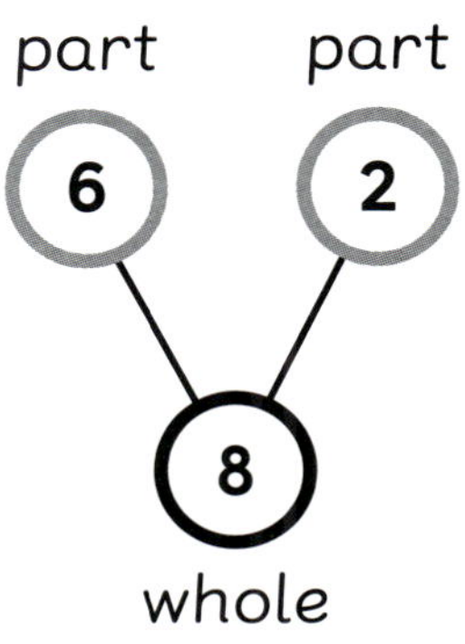

There are 8 hermit crabs altogether.

3

+	0	1	2	3	4	5	6	7	8	9	10
0	0 + 0	0 + 1	0 + 2	0 + 3	0 + 4	0 + 5	0 + 6	0 + 7	0 + 8	0 + 9	0 + 10
1	1 + 0	1 + 1	1 + 2	1 + 3	1 + 4	1 + 5	1 + 6	1 + 7	1 + 8	1 + 9	
2	2 + 0	2 + 1	2 + 2	2 + 3	2 + 4	2 + 5	2 + 6	2 + 7	2 + 8		
3	3 + 0	3 + 1	3 + 2	3 + 3	3 + 4	3 + 5	3 + 6	3 + 7			
4	4 + 0	4 + 1	4 + 2	4 + 3	4 + 4	4 + 5	4 + 6				
5	5 + 0	5 + 1	5 + 2	5 + 3	5 + 4	5 + 5					
6	6 + 0	6 + 1	6 + 2	6 + 3	6 + 4						
7	7 + 0	7 + 1	7 + 2	7 + 3							
8	8 + 0	8 + 1	8 + 2								
9	9 + 0	9 + 1									
10	10 + 0										

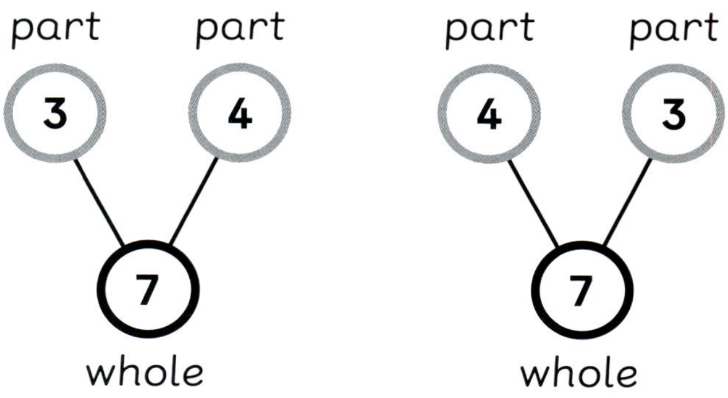

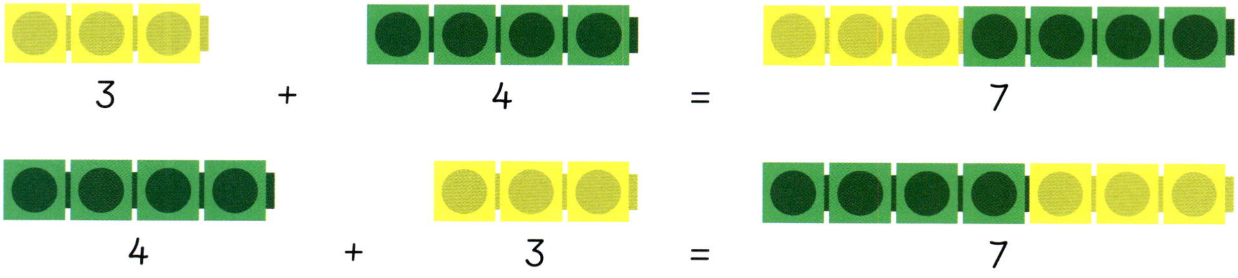

Guided Practice

Complete the number bonds and fill in the blanks.

1

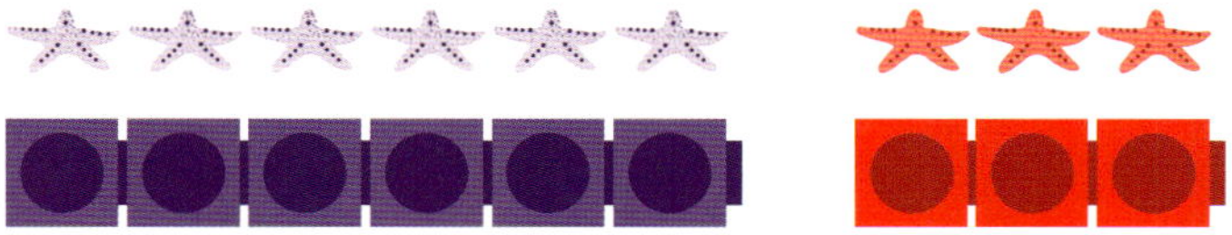

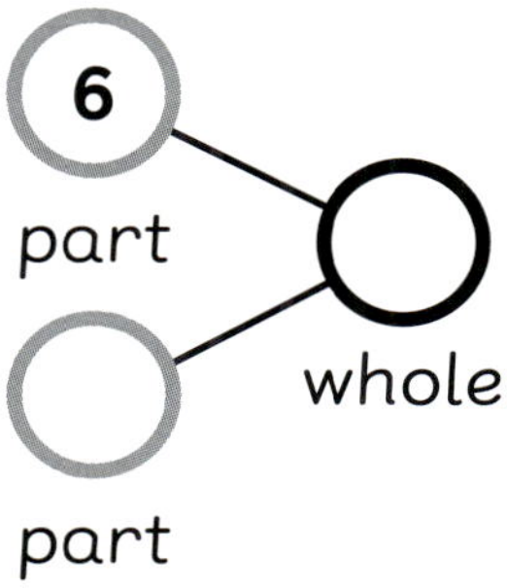

6 + ____ = ____

There are ____ starfish altogether.

2

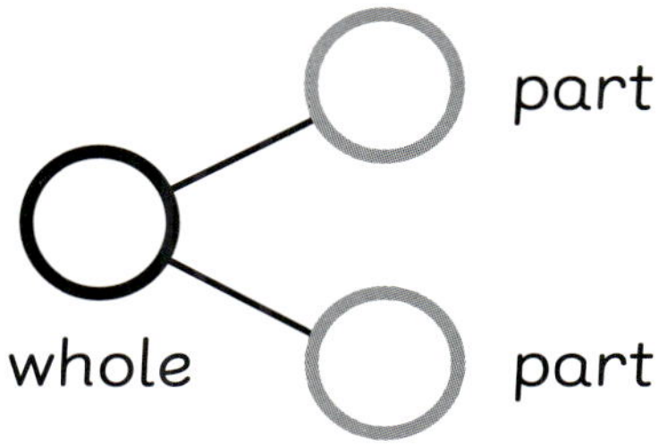

____ + ____ = ____

There are ____ birds altogether.

3

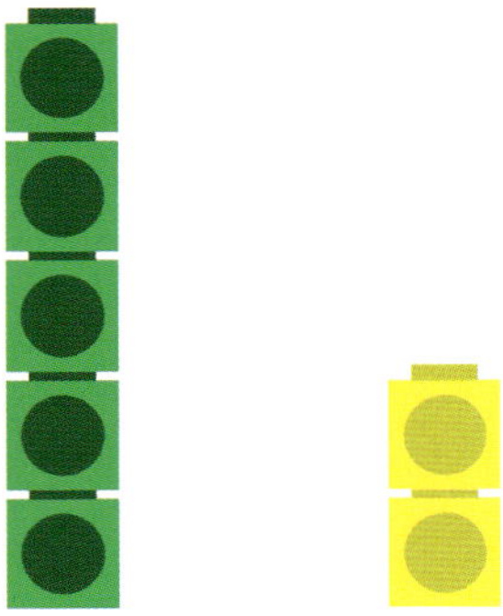

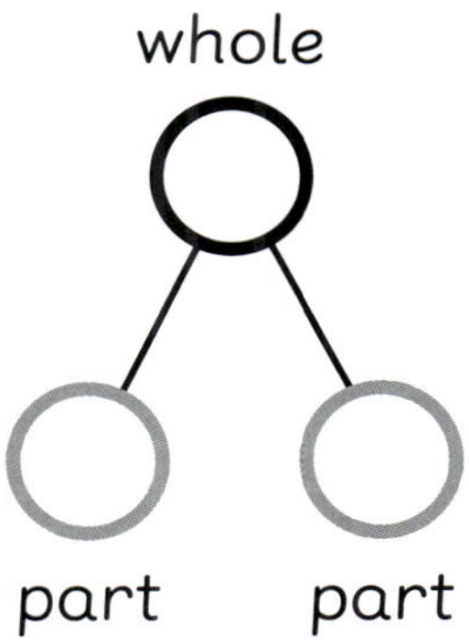

____ + ____ = ____

4

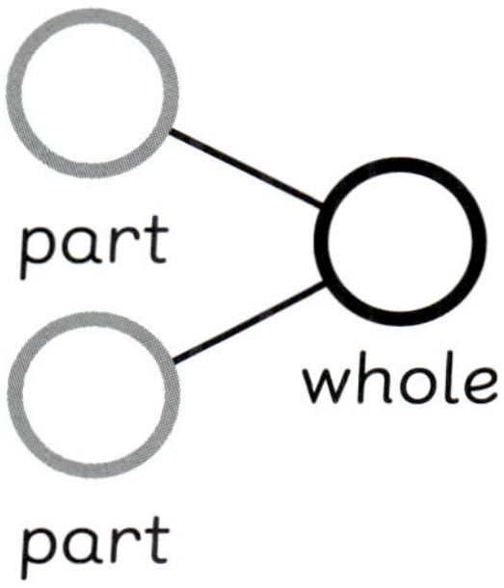

____ + ____ = ____

Complete Worksheet **1**: pages **31–32**

Lesson 2

Add by Counting On

Explore

How many blocks are there altogether?

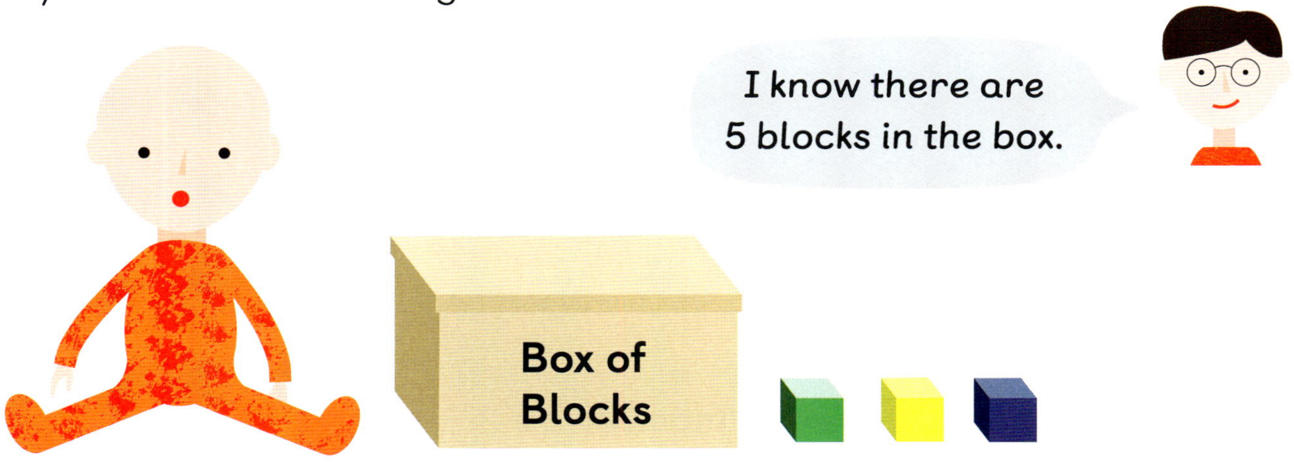

Master

1 5 + 3 = ☐

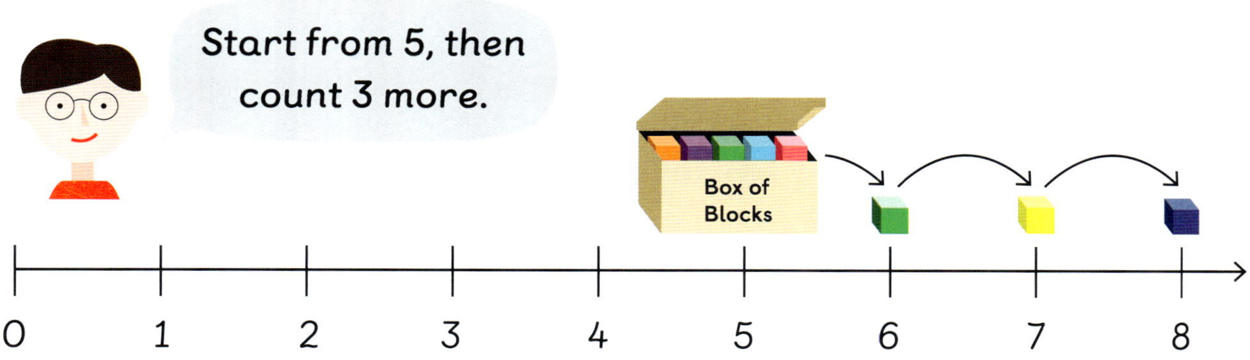

5 + 3 = 8
There are 8 blocks altogether.

2 How many toys are there in total?

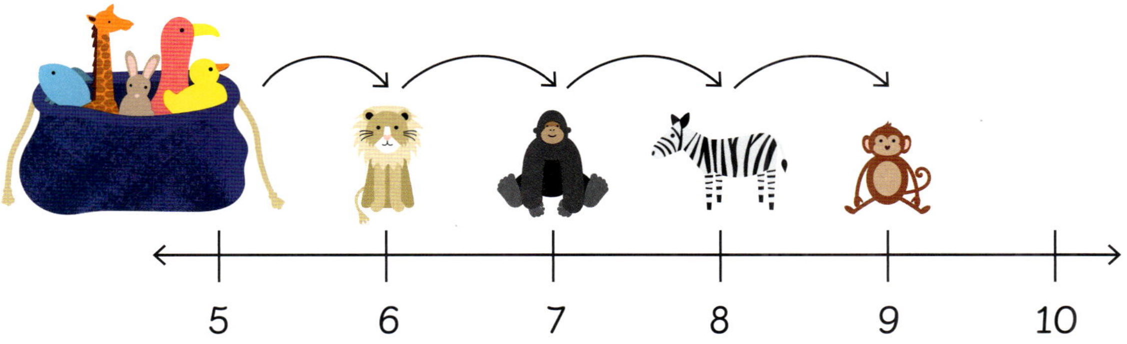

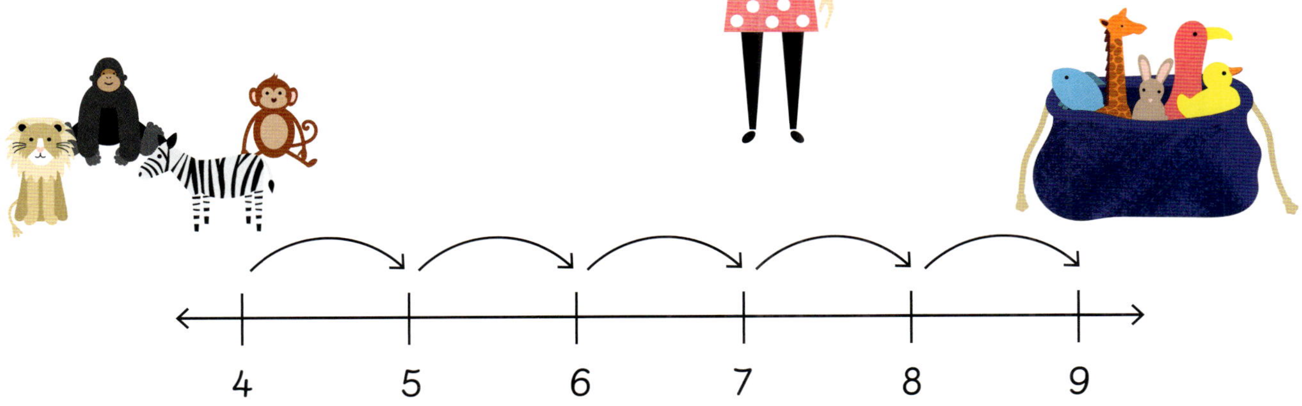

There are ▭ toys in total.

Activity Time

Partner work

You will need:

2  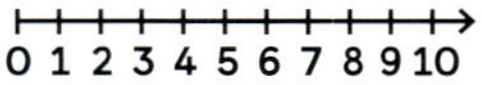

1. Place  face down.
2. Turn over one card each.
3. Add by counting on using the number line.
4. Repeat the activity.

Guided Practice

Count on to find the total number of objects.

1

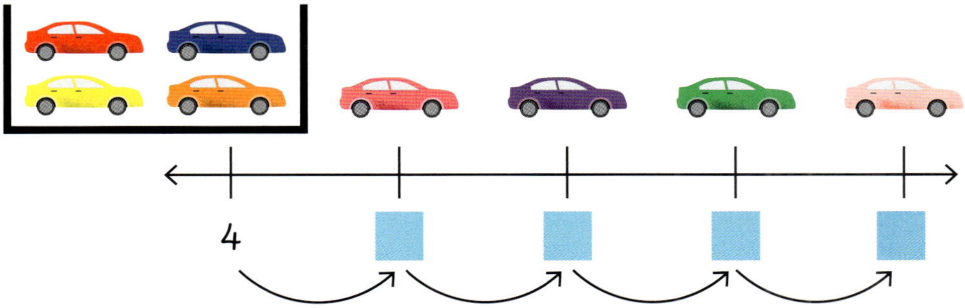

4 + 4 = ____

There are ____ cars in total.

2

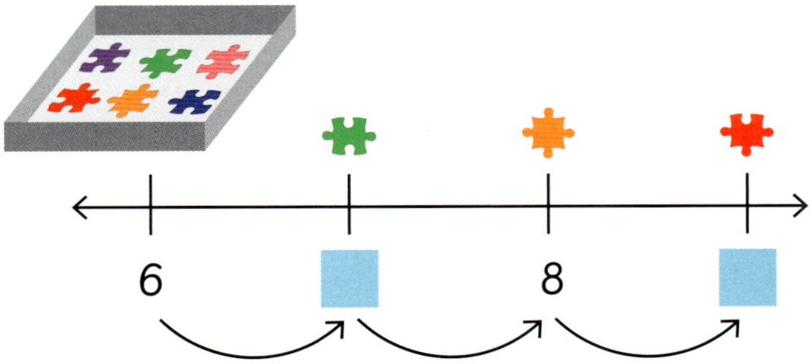

6 + ____ = ____

There are ____ puzzle pieces altogether.

3

____ + ____ = ____

There are ____ balls in total.

Complete Worksheet **2**: pages **33–36**

Completing Number Sentences

Lesson 3

Explore

Lulu has 8 coins.
How many are in the bag?

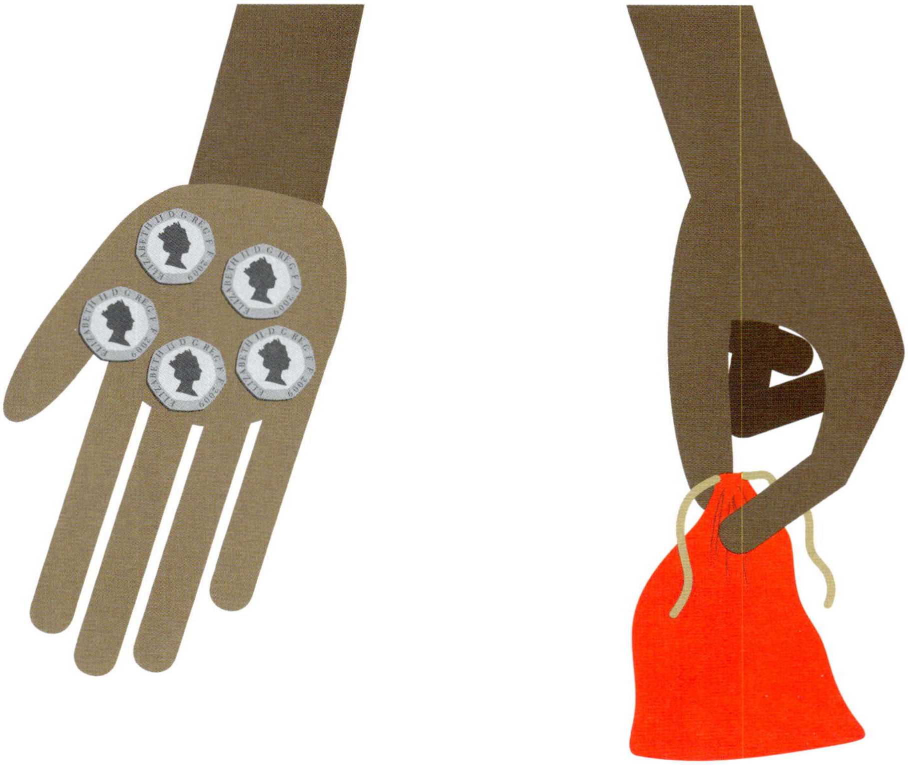

Master

1 5 + ☐ = 8

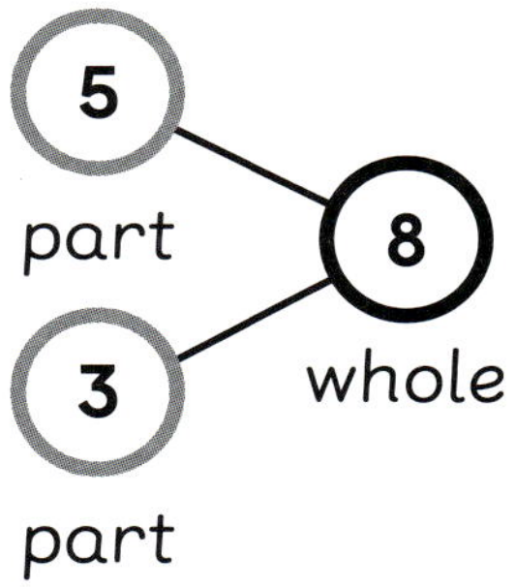

2

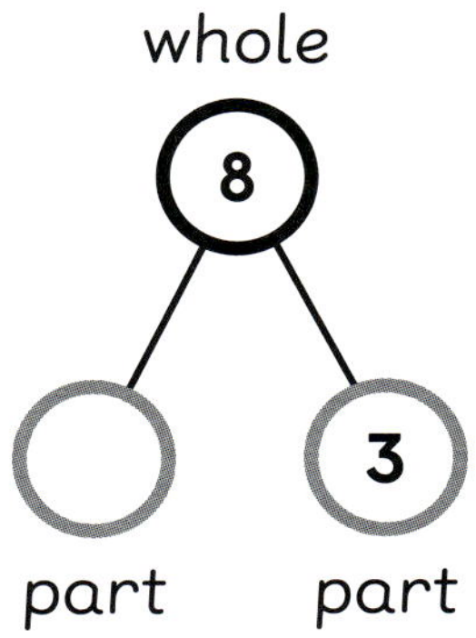

8 = ☐ + 3

There are 3 coins in the bag.

Guided Practice

1

6 and ___ is equal to 9.

6 + ___ = 9

9 = 6 + ___

2 Find another way to make 9.

___ + ___ = 9

3

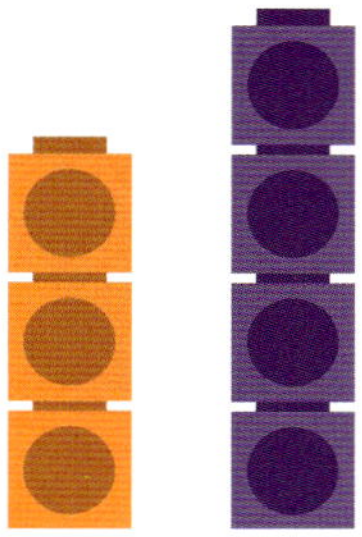

whole

part part

☐ + ☐ = ☐

4

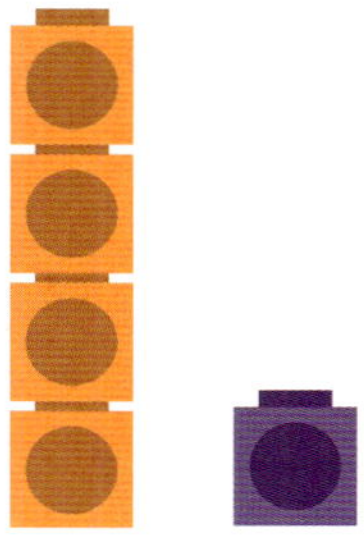

whole

part part

☐ + ☐ = ☐

5

whole

7

4

part part

☐ + ☐ = ☐

Complete Worksheet **3**: pages **37–38**

Making Addition Stories

Lesson
4

Explore

There are 4 adult seals and 1 baby seal.
There are 5 seals in total.

Can you make your own addition story?

Master

1 There are 4  on the sand.

There are 3  on the rocks.

4 + 3 = 7

There are 7 puffins altogether.

2 There are 8  flying.

There are ___  on the rocks.

8 + ___ = ___

There are ___ seagulls in total.

3 There are 3  on the sand.

There are ___ 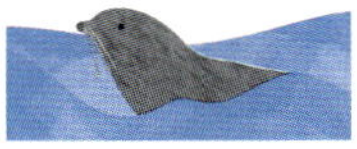 swimming.

3 + ___ = ___

There are ___ seals altogether.

Guided Practice

Fill in the blanks.

1 There are ☐ children in the sea.

There are ☐ children on the beach.

☐ + ☐ = ☐

There are ☐ children in total.

2 There are ☐ birds on the rocks.

☐ birds are on the beach.

☐ + ☐ = ☐

☐ birds are not flying.

Complete Worksheet **4**: pages **39–42**

Solving Picture Problems

Lesson 5

Explore

How many hula hoops does Hannah have in total?

Master

1

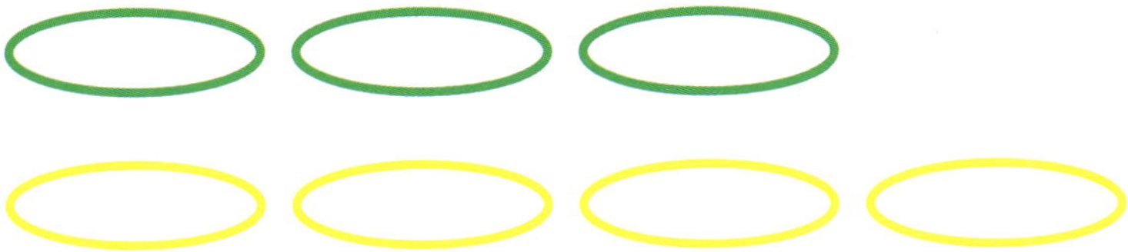

3 + 4 = 7

Hannah has 7 hula hoops in total.

2

4 + 5 = ☐

There are ☐ books altogether.

Guided Practice

1 Count the footballs.

There are ☐ blue footballs.

There are ☐ white footballs.

☐ + ☐ = ☐

There are ☐ footballs altogether.

2

How many pies are there in total?

☐ + ☐ = ☐

There are ☐ pies in total.

Complete Worksheet **5**: pages **43–46**

Mind Challenge

Use the digit cards to complete the square.

7 4 3 1 0 2

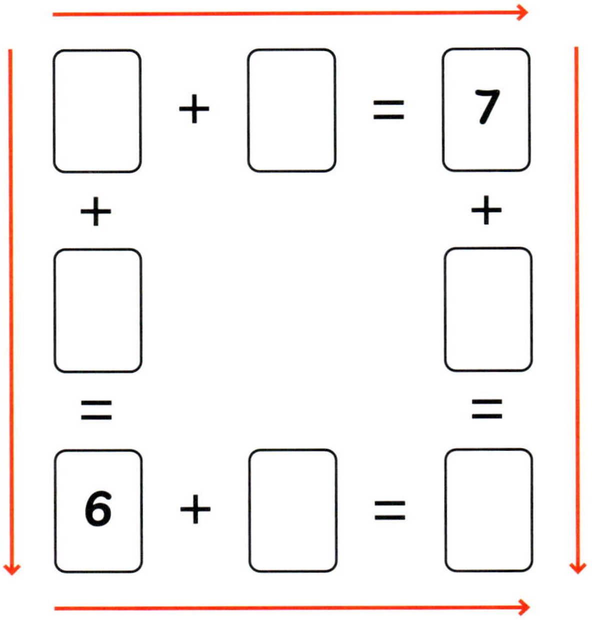

My Maths Journal

☐ + ☐ = 5

How many ways can you solve this?

You can use pictures and words.

Review

I can...

- ☐ add by counting on.
- ☐ add by using number bonds.
- ☐ make and tell addition stories.
- ☐ write and solve addition picture problems.

Chapter 4
Subtraction within 10

Subtract by Crossing Out

Lesson 1

Explore

There are 8 monkeys swinging in a tree.
3 jump off.
How many monkeys are left in the tree?

Master

1

8 − 3 = 5

5 monkeys are left in the tree.

8 − 3 = 5 is a subtraction equation.
We say eight minus three equals five.

2 There are 7 snakes. 4 go to hide.
How many snakes are not hiding?

7 − 4 = 3

There are 3 snakes not hiding.

Guided Practice

Use crossing out to help you subtract.

1 $6 - 2 =$ ☐

2 $9 - 4 =$ ☐

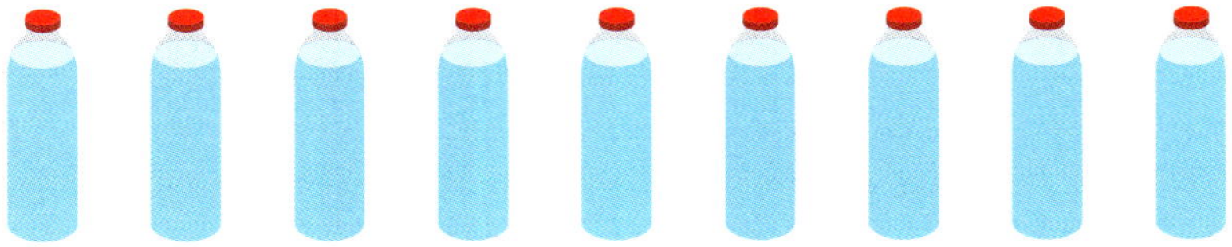

3 $8 - 4 =$ ☐

Complete Worksheet **1**: pages **53–54**

Subtract by Using Number Bonds

Lesson 2

Explore

How many elephants are not adults?

Master

1

6 − 4 = 2

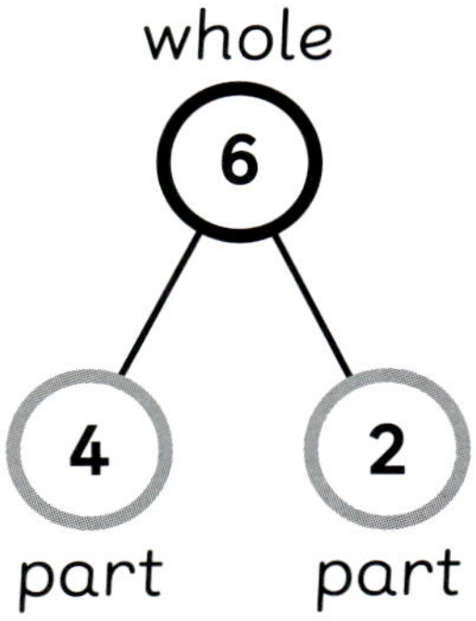

2 elephants are not adults.

2 There are 8 cakes in all.

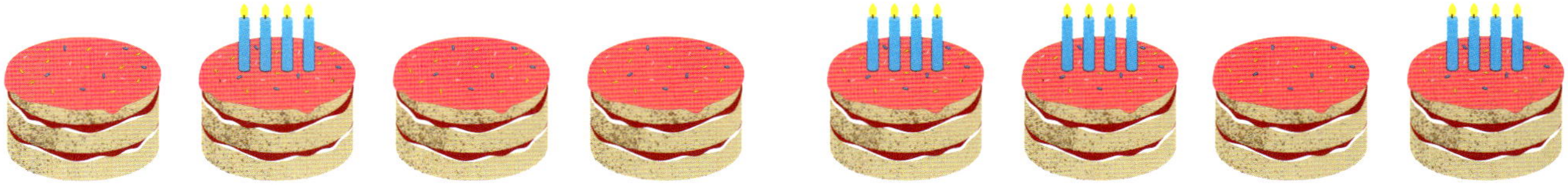

How many cakes have candles?

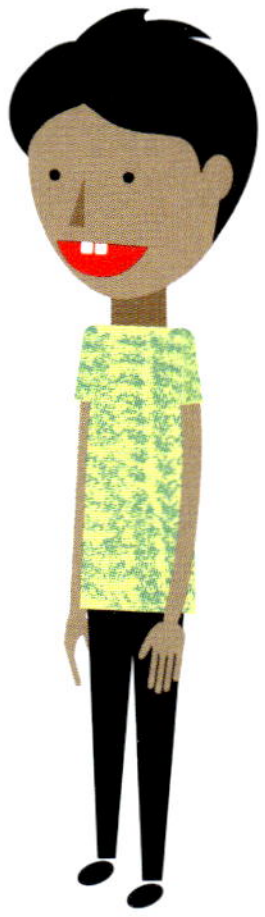

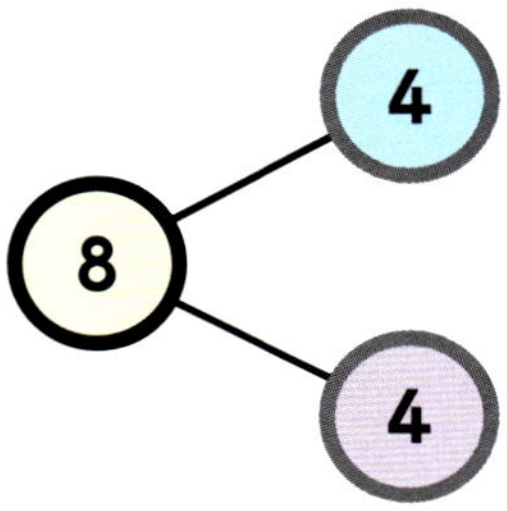

8 − 4 = 4
whole part part

There are 4 cakes with candles.

Guided Practice

Use number bonds to help you subtract.

1

6 − 2 = ___

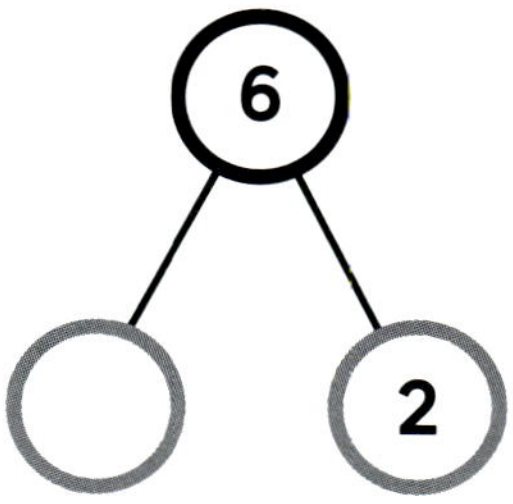

There are 6 children. 2 are wearing hats.

There are ___ children not wearing hats.

2

7 − 5 = ▭

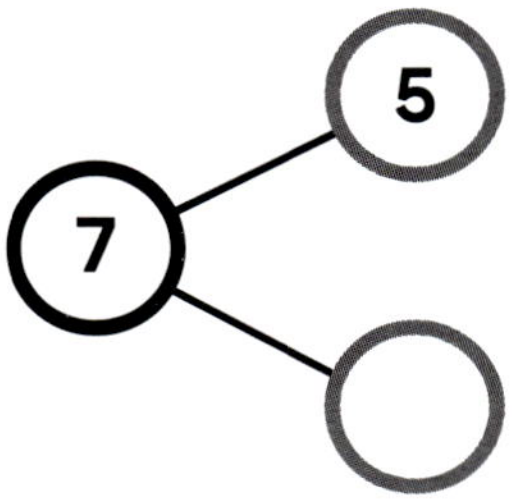

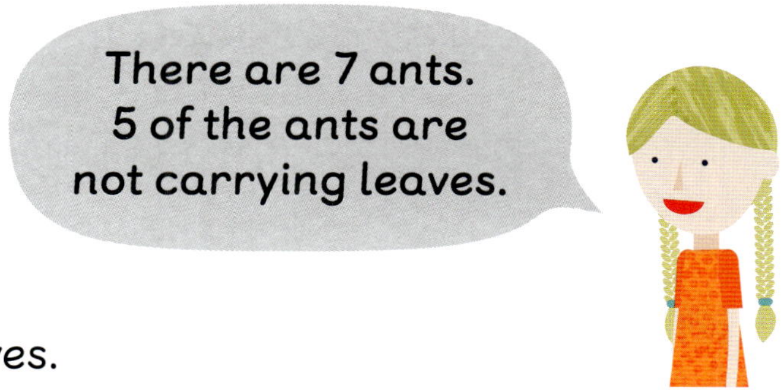

There are ▭ ants carrying leaves.

3

10 − 6 = ▭

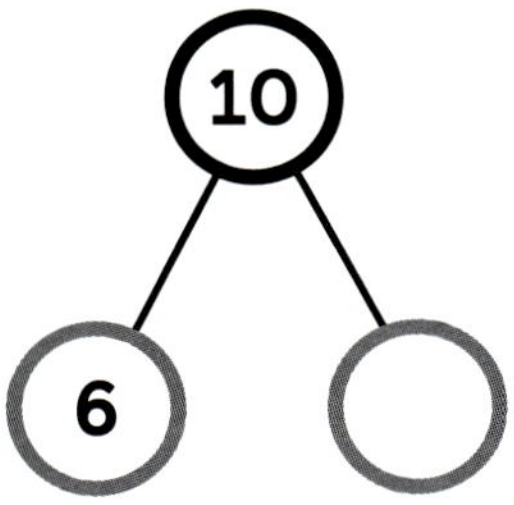

There are ▭ leopards lying in the tree.

Complete Worksheet 2: pages 55–58

Lesson 3

Subtract by Counting Back

Explore

There are 7 carrots altogether.
Sam takes 3 carrots.
How many carrots are left in the bag?

Master

1

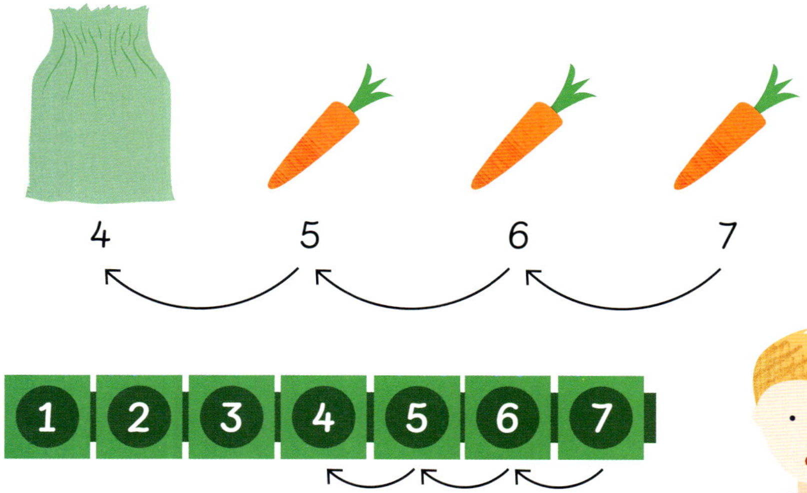

7 – 3 = 4

There are 4 carrots left in the bag.

2 6 friends are playing together.
2 of the friends are playing outside.
How many friends are playing in the house?

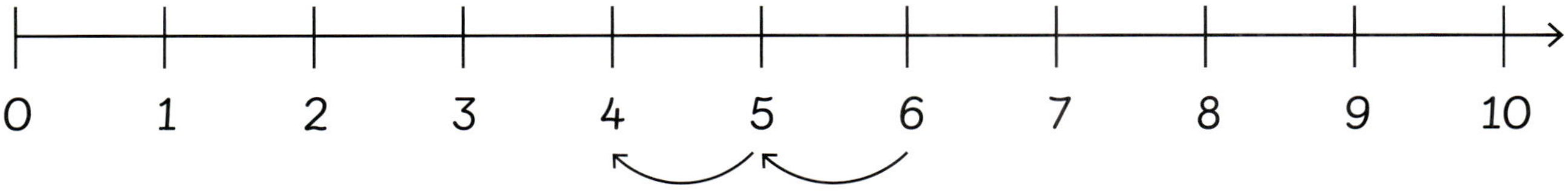

6 − 2 = 4

There are ☐ friends playing in the house.

Guided Practice

Solve.

1 Sam's mum bought 5 mangoes in total.
How many of the mangoes are in the bag?

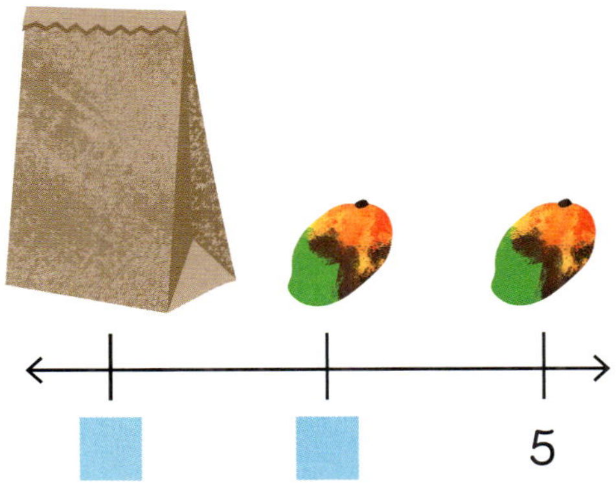

5 − 2 = ☐

There are ☐ mangoes in the bag.

2 Amira has 7 pictures altogether.
How many pictures are there in her photo album?

7 − ___ = ___

There are ___ pictures in her photo album.

3 Holly has 10 pencils in total.
How many pencils are in her pencil case?

10 − ___ = ___

Complete Worksheet **3**: pages **59–60**

Making Subtraction Stories

Lesson 4

Explore

Master

1

6 − 2 = 4

4 animals are not drinking.

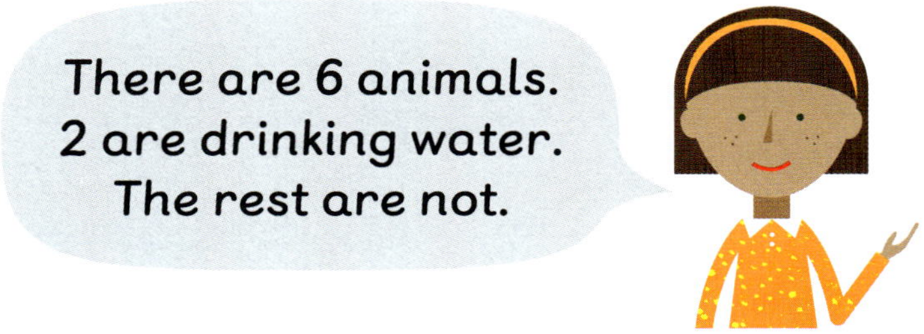

2 Before the animals arrived, there were 8 leaves on the trees.

The animals ate 3 leaves.

8 − 3 = 5

5 leaves were left on the trees.

Activity Time

Partner work

You will need:

1. Pick a  .
2. Draw a subtraction picture to match the card.
3. Your partner tells the subtraction story.
4. Take turns.

Guided Practice

Complete the subtraction stories.

1 There are 8 rhinoceros.

How many are white? [ ]

How many are black?

8 − [ ] = [ ]

There are [ ] black rhinoceros.

2 There are [ ] birds.

[ ] have landed on the rhinoceros.

[ ] − [ ] = [ ]

[ ] have not landed on the rhinoceros.

Complete Worksheet **4**: pages **61–62**

Solving Picture Problems

Lesson 5

Explore

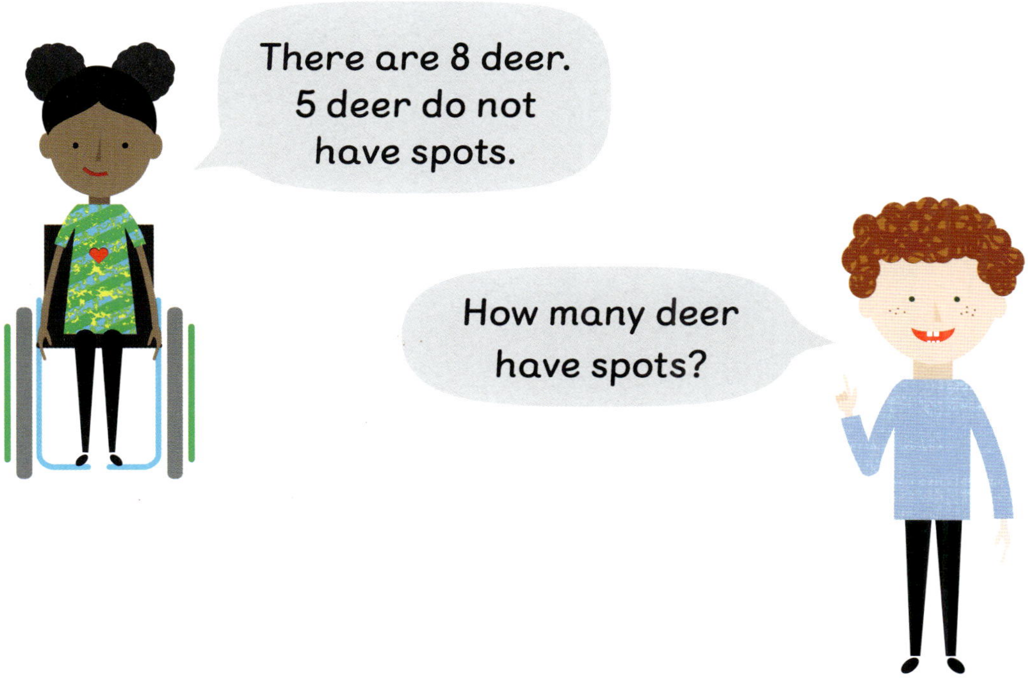

Master

1

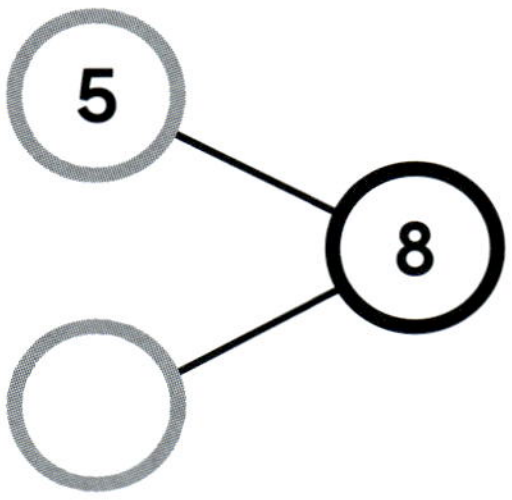

$8 - 5 = 3$

There are 3 deer with spots.

2

There are 10 pigeons flying.

(a) How many pigeons are white?

10 − ___ = ___

___ pigeons are white.

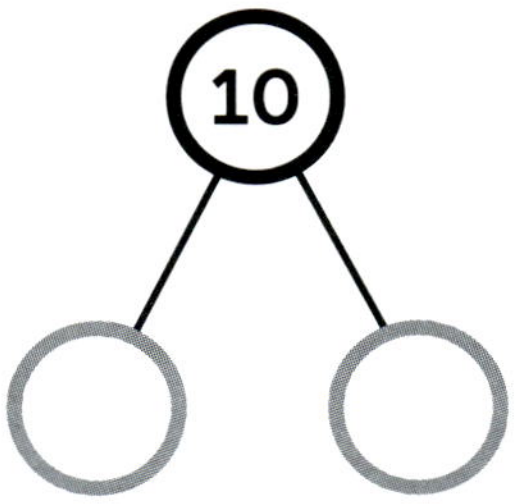

(b) ___ − ___ = ___

___ pigeons are grey.

Guided Practice

Fill in the blanks.

1

There are 7 toys.

☐ toys are red.

7 − ☐ = ☐

☐ toys are blue.

2

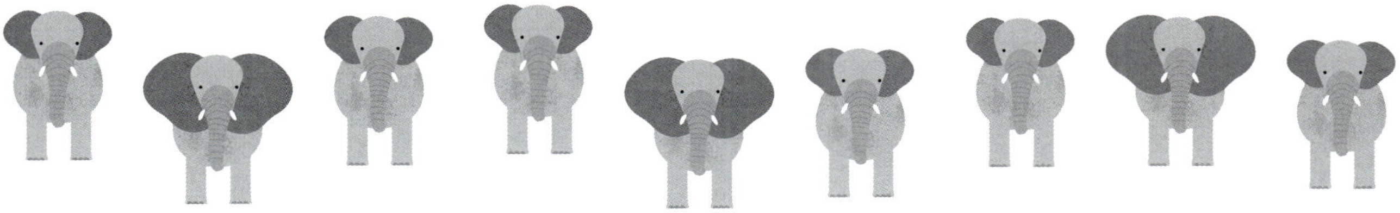

If the elephants with small ears walk away, how many elephants are left?

☐ − ☐ = ☐

☐ elephants are left.

Complete Worksheet **5**: pages **63–64**

Addition and Subtraction

Lesson 6

Explore

What addition and subtraction stories can you tell?

Master

There are 10 bears.
3 bears are black and
7 bears are brown.

There are 10 bears altogether.
3 + 7 = 10   or   7 + 3 = 10

There are 3 black bears.
10 − 7 = 3

There are 7 brown bears.
10 − 3 = 7

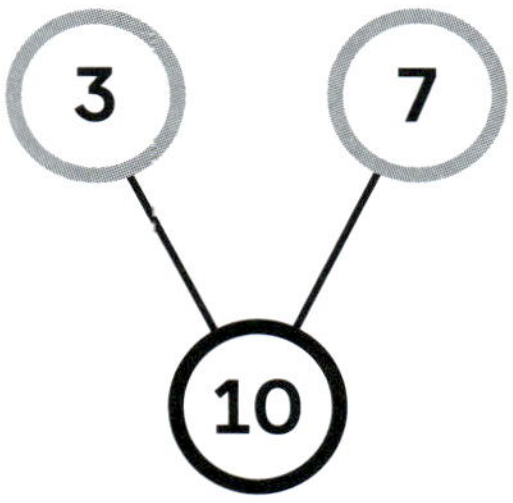

These equations make up an addition and subtraction fact family.

Activity Time

Play a game (2 pupils)

You will need: 4 + 5 | 9 − 1

1. Lay all the cards face up on the table.

2. Take turns to match an addition fact to a subtraction fact from the same fact family.

   For example:

   You pick 4 + 5.

   If your partner picks 9 − 5, that makes a pair!

3. Play until all the cards are in pairs.

Guided Practice

Look at the picture.
Make an addition and subtraction fact family.

☐ + ☐ = ☐   ☐ − ☐ = ☐

☐ + ☐ = ☐   ☐ − ☐ = ☐

Complete Worksheet 6: pages 65–66

Mind Challenge

Use all the number facts you know to complete the diagram.

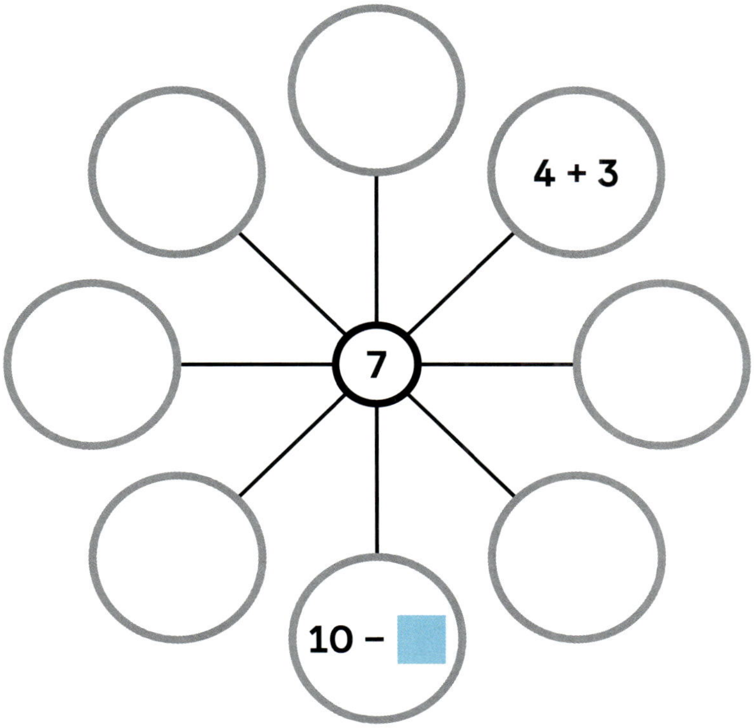

My Maths Journal

Use the picture to help you write an addition story and a subtraction story.
How would you solve your addition story?
How would you solve your subtraction story?

I can...

Review

- ☐ use crossing out to subtract.
- ☐ use number bonds to subtract.
- ☐ count back to subtract.
- ☐ make stories using subtraction.
- ☐ write subtraction equations.
- ☐ make addition and subtraction fact families.

In what position is each boat from the finishing line?

Chapter 5
Positions

Naming Positions

Lesson 1

Explore

Is Lulu correct?

Master

1st	2nd	3rd	4th	5th
first	second	third	fourth	fifth

The  came first. We can also say the  won the race.

The  came second and the  came third.

Lulu is correct. The  came fourth.

The  came fifth. We can also say the  came last.

The  came after the  but before the  .

Activity Time

Group work (4 pupils)

1. Put the 

in order.

2. Put the 

in the correct order above the 

.

3. Read each action and its position.

You will need:

Guided Practice

1. The ________ came first in the race.
2. The fish came ________ in the race.
3. The ________ came fourth in the race.
4. The fish came before the ________ and after the ________.
5. Which animal came last? ________

Complete Worksheet 1: pages 79–82

Naming Positions in Queues

Lesson 2

Explore

Name the positions of the animals queuing to drink at the waterhole.

Master

The  is 1st in the queue.

The  is 2nd.

The  is 3rd.

The  is 4th.

The  is 5th.

The  is between the  and .

The  is before the  and after the .

Guided Practice

The animals are queuing for apples.

1 Name the position of each animal.

chicken	sheep	cow	goat	horse
		3rd		
		third		

2 The  is ______ in the queue.

3 The ______ is third in the queue.

4 The ______ is last in the queue.

5 The  is between the ______ and the ______.

6 The animal before the  and after the  is the ______.

Complete Worksheet **2**: pages **83–84**

Naming Left and Right Positions

Lesson 3

Explore

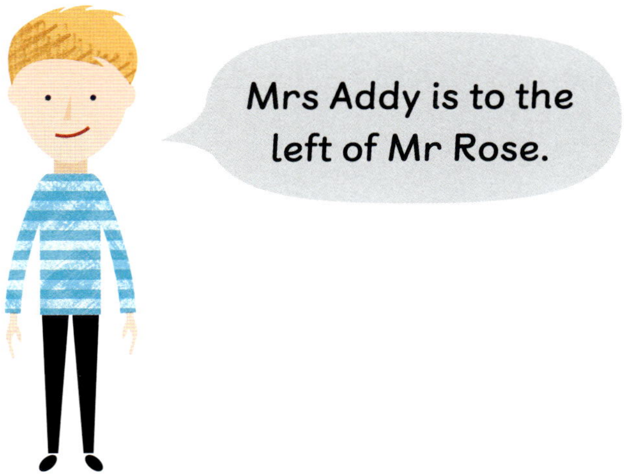

Who is correct?

Master

1

We can name the positions of the adults from the left or the right.

 is first from the left.

 is second from the right.

 is fourth from the left and first from the right.

 is between  and .

 is next to . He is also next to .

Sam and Jacob are both correct!

2

In the first row,  is third from the left.

She is also first from the right.

In the second row,  is first from the left.

She is also fourth from the right.

Guided Practice

left  right

red boat	green boat	yellow boat	blue boat	pink boat	orange boat

Complete these sentences.

1 The ________ boat is third from the left.

2 The  is ________ from the right.

It is also ________ from the left.

3 The  is next to the ________ boat.

4 The  is between the ________ boat and the ________ boat.

Complete Worksheet **3**: pages **85–86**

Mind Challenge

Boats are queuing to get into the dock.
The red boat is third from the end of the queue.
How many boats are in the queue altogether?

My Maths Journal

Describe the positions of the cars.
Use words such as left, right, before, next to, between, first, second and third.

Review

I can...

- ☐ name positions in a race.
- ☐ name positions in a queue.
- ☐ name positions from left to right and right to left.
- ☐ use the words before, after, next to, between and last to describe positions.

Chapter 6
Numbers to 20

Counting to 20

Lesson 1

Explore

How many plants are there in total?

Master

1

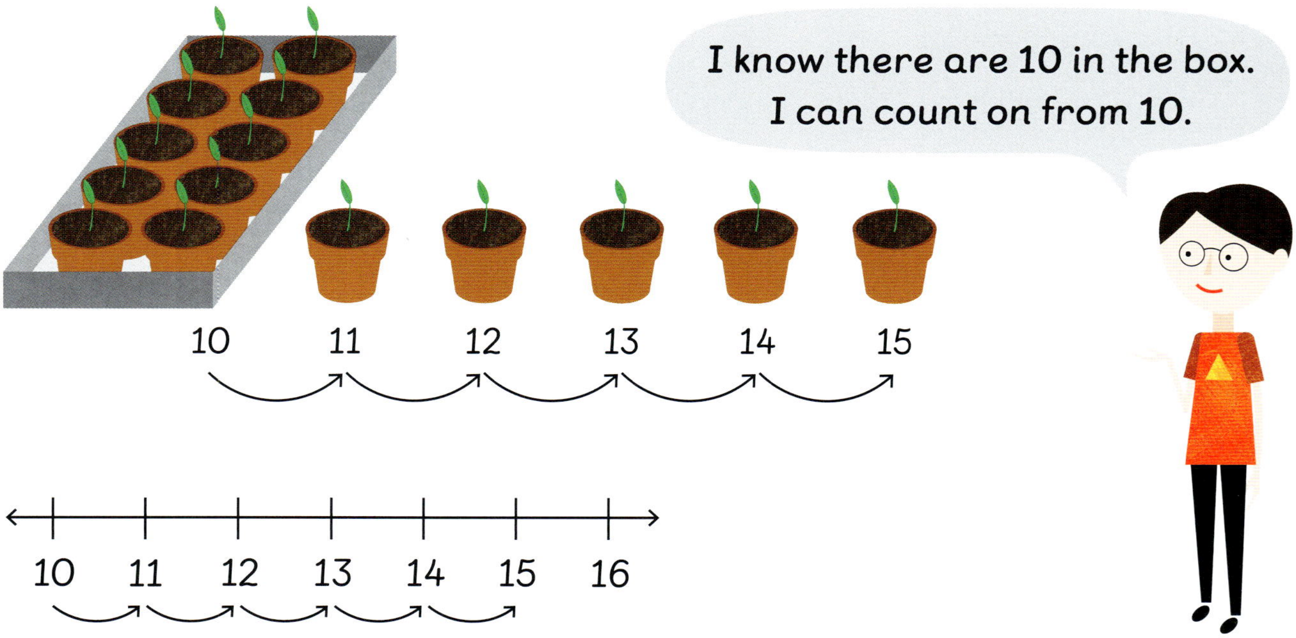

There are 15 plants in total.

2 How many bees can you count?

10 + 7 = 17

There are 17 bees in total.

3

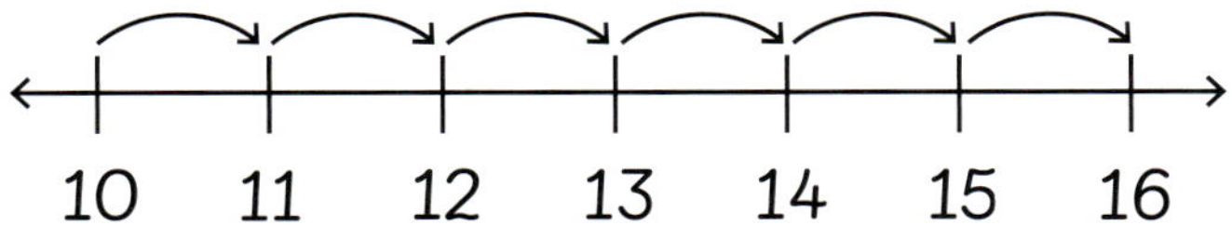

10 + 6 = 16

There are 16 pencils.

3 Help Amira by counting on from 10.

11

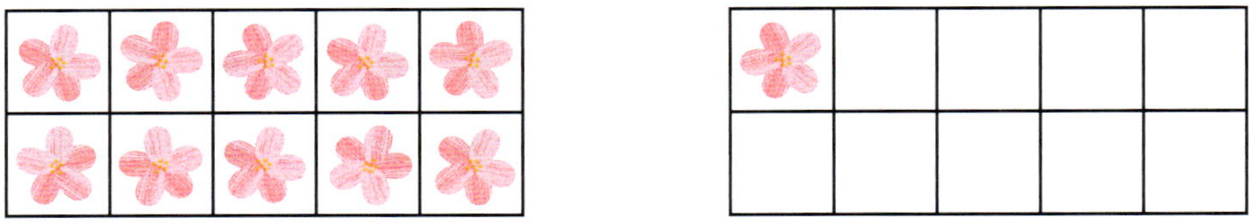

12

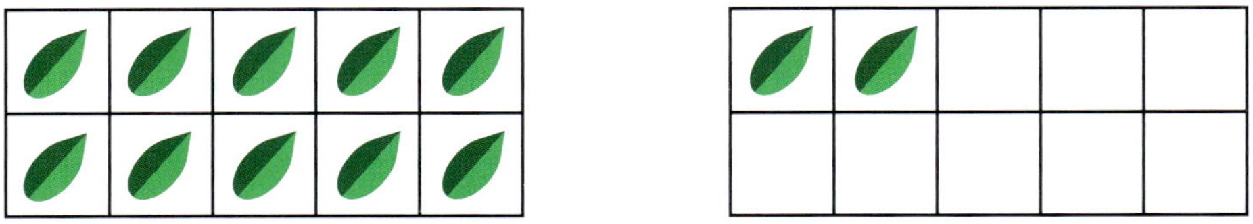

13

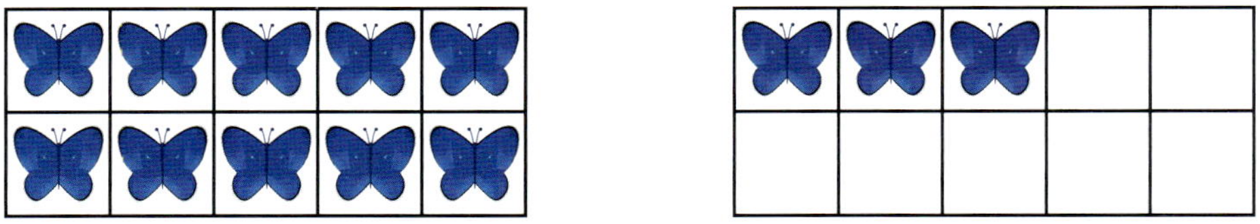

14

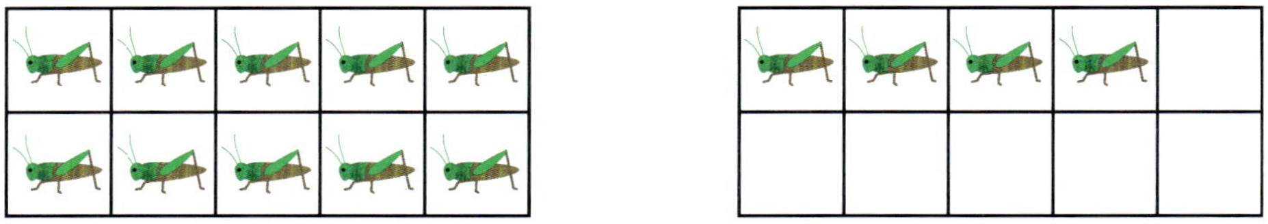

15

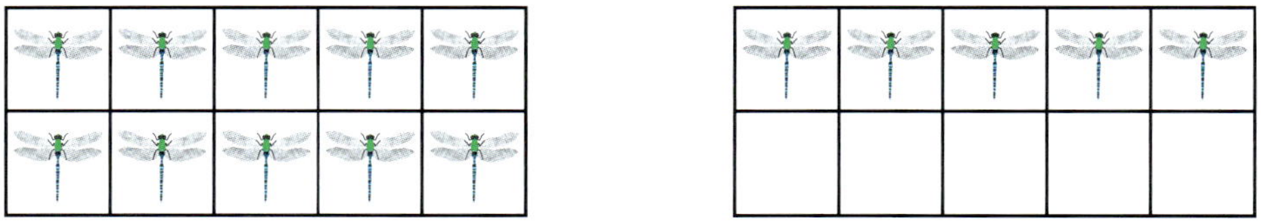

16

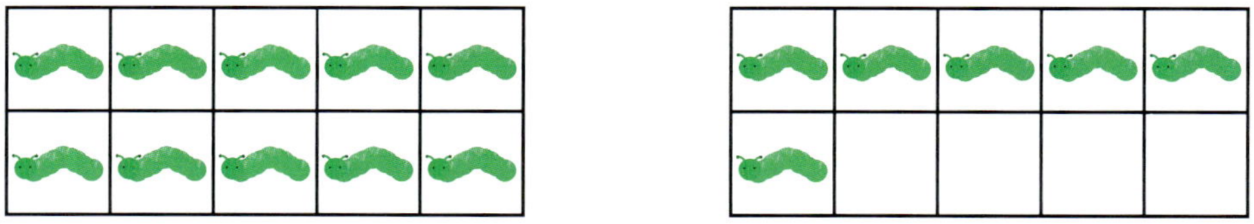

17

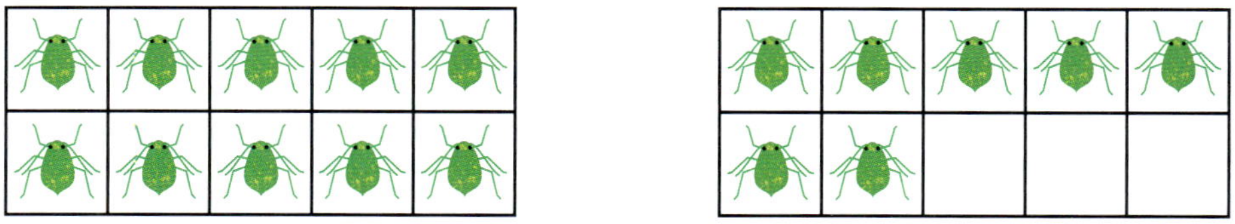

18 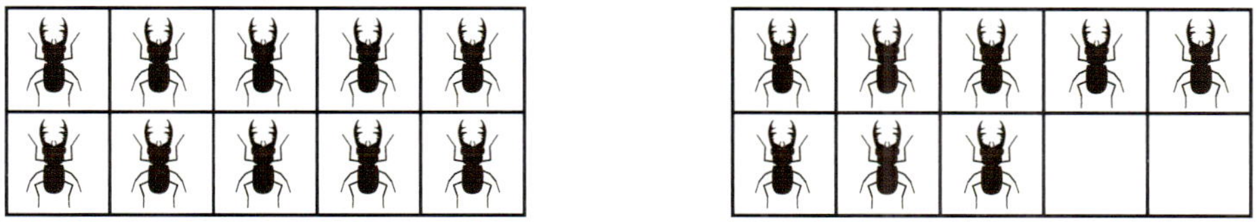

19 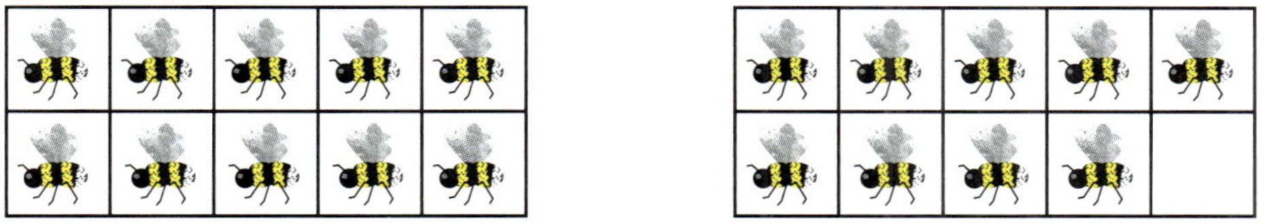

20 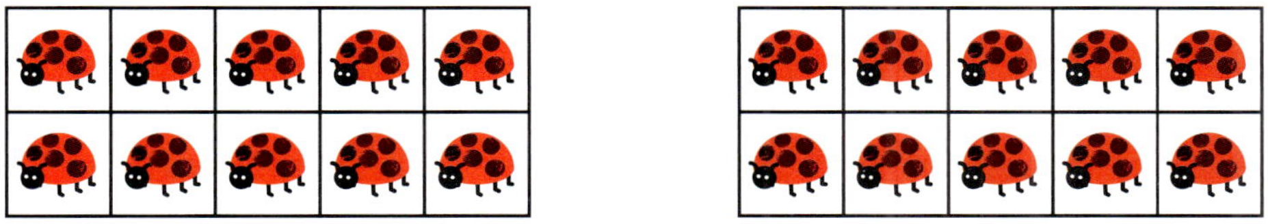

Activity Time

Partner work

You will need:

 20  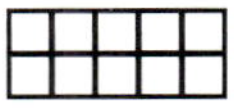

1. Put 10  on a 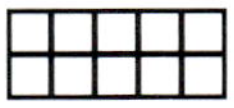.
2. Ask your partner to pick a card and place the same number of  beside 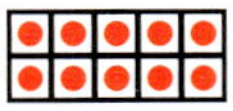.

   For example:
   If your partner picks 3, they place 3  beside 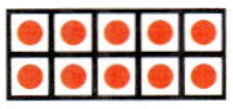.
3. Together, work out what the new number is by counting on from 10.
4. Take turns to write the addition equation.

   For example: 10 + 3 = 13

Guided Practice

1

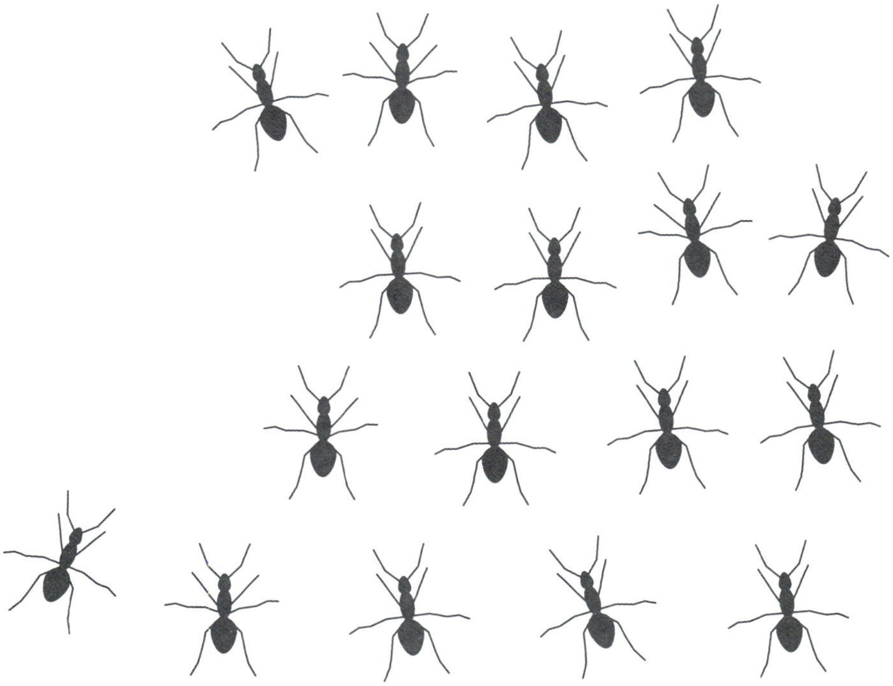

10 + 7 = ____

There are ____ ants.

2

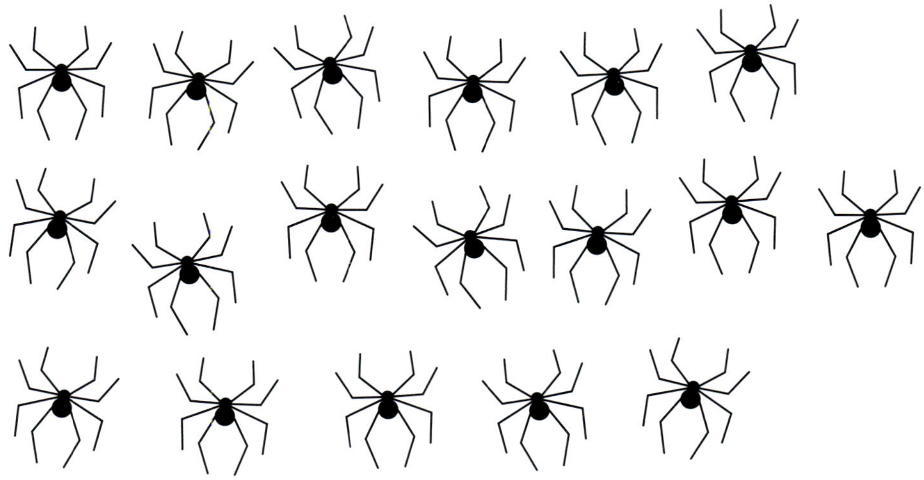

____ + ____ = ____

There are ____ spiders.

Complete Worksheet **1**: pages **91–92**

Writing to 20

Lesson 2

Explore

How many apples are there?
Can you write the number in words?

Master

1

There are 12 .

2 , ,  and  take a walk and see lots of different flowers.

eleven  11

twelve 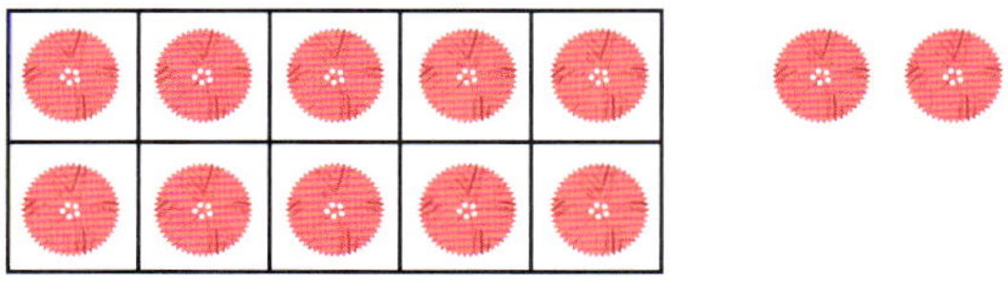 12

thirteen 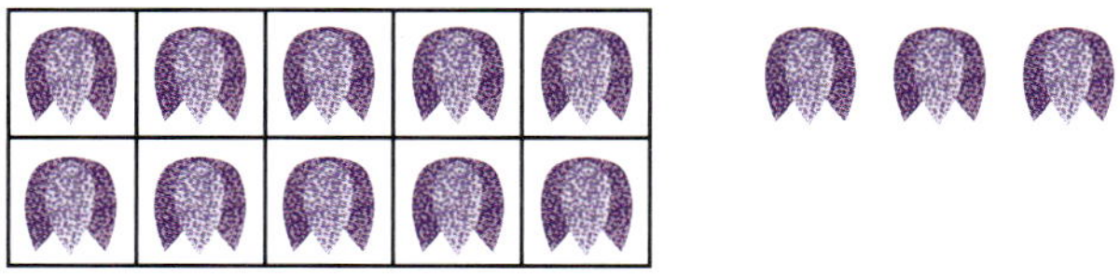 13

fourteen 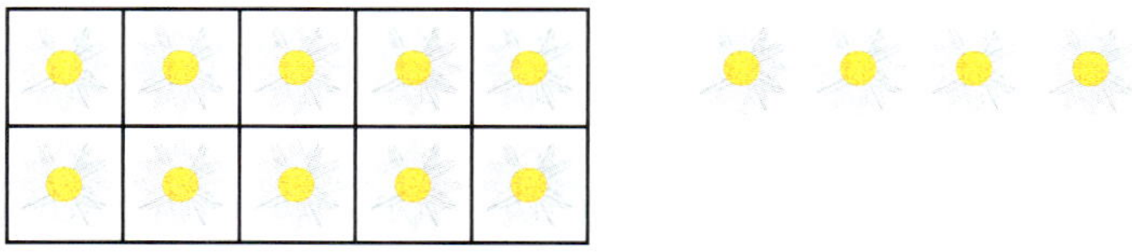 14

fifteen  15

sixteen  16

seventeen 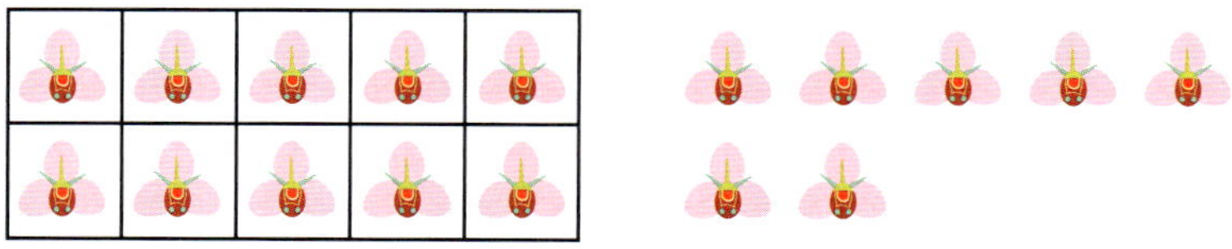 17

eighteen  18

nineteen 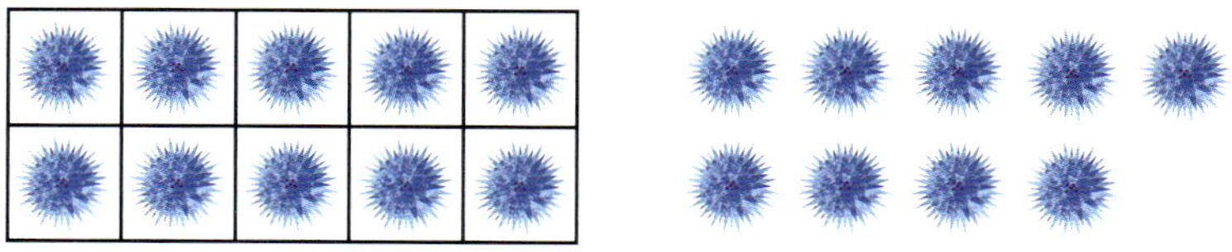 19

twenty  20

3

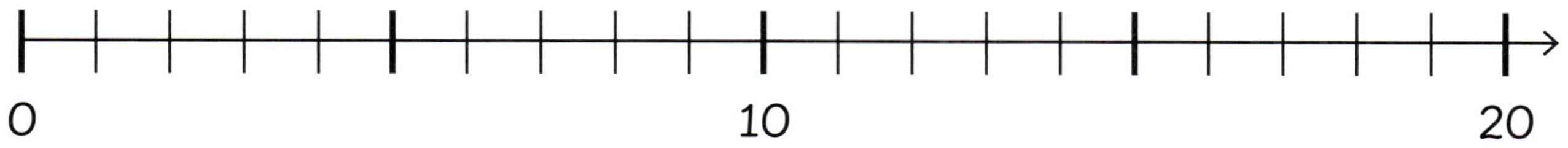

Activity Time

Partner work

You will need:

| 1 twenty | 20 nineteen | 19 eighteen | 18 seventeen |

1. Shuffle the loop cards.
2. Take turns to place a card down to match a word to a number.
3. Continue until you have used all the cards.

Guided Practice

1 Make 10 and count on.
Write the amount in numbers and in words.

(a)

(b)

2 Fill in the missing numbers.

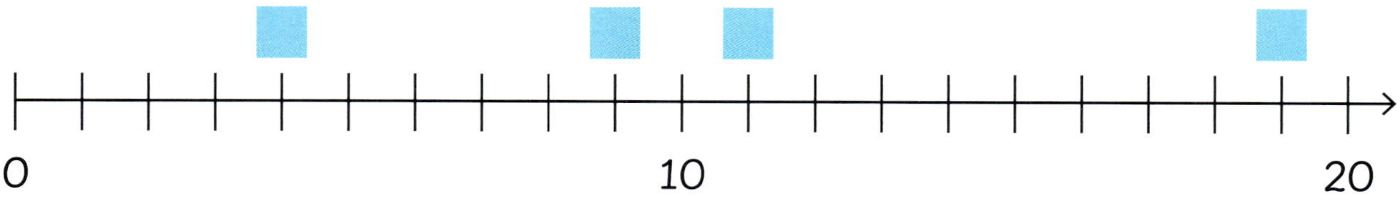

Complete Worksheet **2**: pages **93–94**

Comparing Numbers

Lesson 3

Explore

Which team has more players?

Master

Team A
12 players

Team B
15 players

Team B has **more** players than Team A.

Team A has **fewer** players than Team B.

Activity Time

Group work (2 pairs)

You will need:

20  20 

1. Pair A make a number with 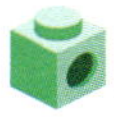.
2. Pair B make a number with .
3. Compare the number of 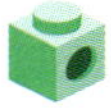 and .
4. Which pair has more? How many more?
5. Which pair has fewer? How many fewer?

Guided Practice

1 Use **more** or **fewer** to compare the objects.

(a) 13 oak leaves

16 sycamore leaves

There are ______  than .

There are ______  than .

(b)

There are ______  than .

There are ______  than .

2 Count and compare the medals.

Group A

There are ____ medals.

Group B

There are ____ medals.

There are ____  than .

There are ____  than .

Complete Worksheet 3: pages 95–96

Ordering Numbers

Lesson 4

Explore

Ravi is using wooden blocks to make a pattern.
How many blocks will he use next?

Master

1

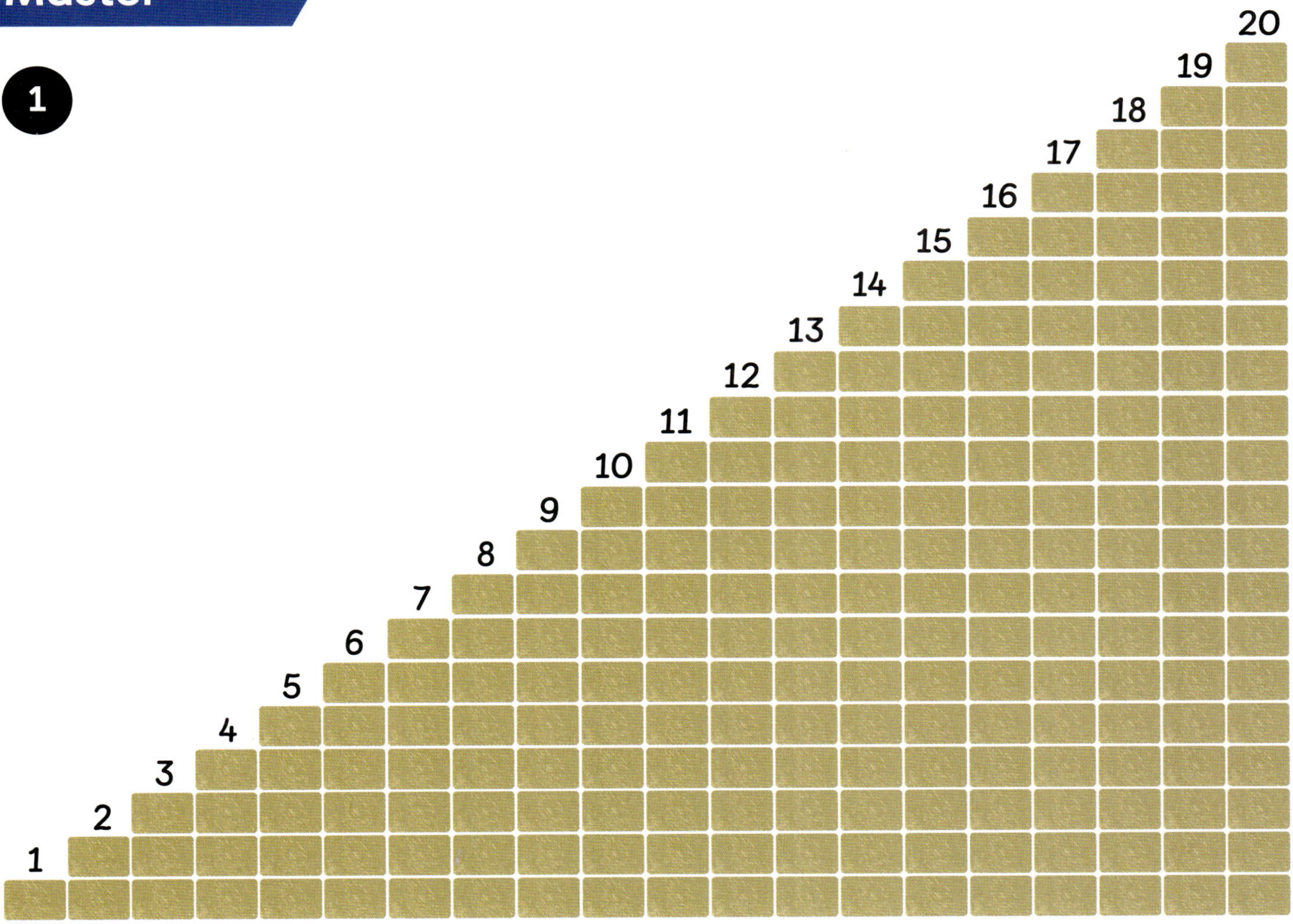

2

2
3
9
10
19
20
3 is 1 more than 2.
10 is 1 more than 9.
20 is 1 more than 19.

3 The numbers can be ordered from smallest to greatest.

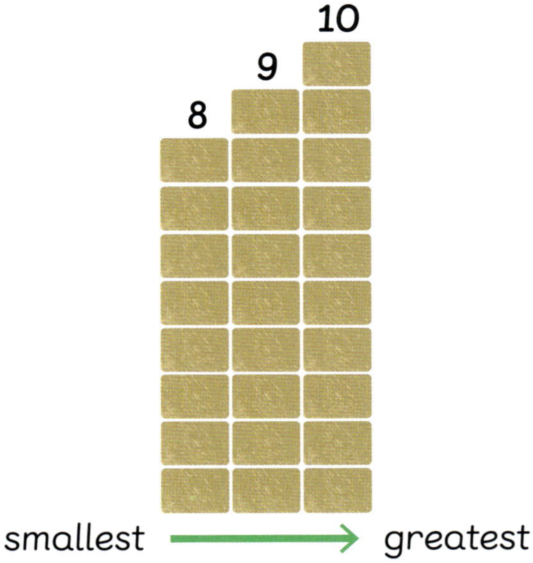

4 The numbers can also be ordered from greatest to smallest.

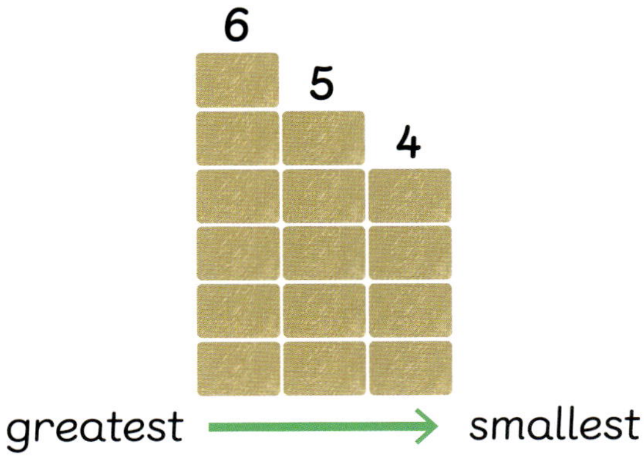

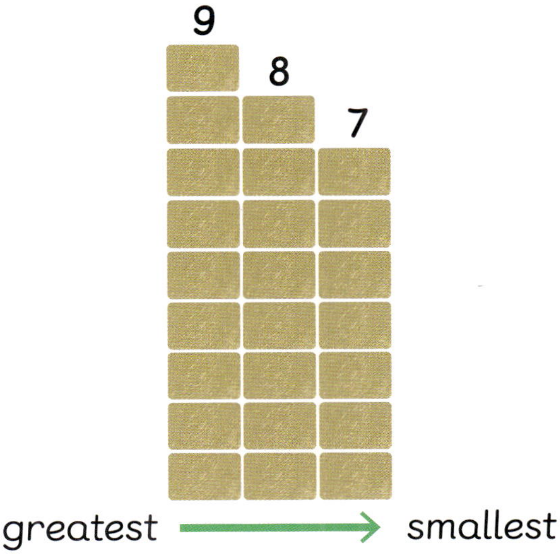

5

Activity Time

Partner work

You will need:

1. Take one card each from the pile.
2. Compare the two numbers.
   Who has the greater number?
   Who has the smaller number?
3. Take another card each from the pile.
4. Order the four numbers from smallest to greatest and then from greatest to smallest.
5. Repeat the activity.

Guided Practice

1 Which is the greatest number?

6	5	8

2 Which is the smallest number?

7	9	8

3 Put the numbers in order from smallest to greatest.

18	8	12

, ,

4 Put the numbers in order from greatest to smallest.

7	16	15

, ,

Complete Worksheet 4: pages **97–98**

Number Patterns

Lesson 5

Explore

12   13   14

How many drinks will be in the next group?

Master

1

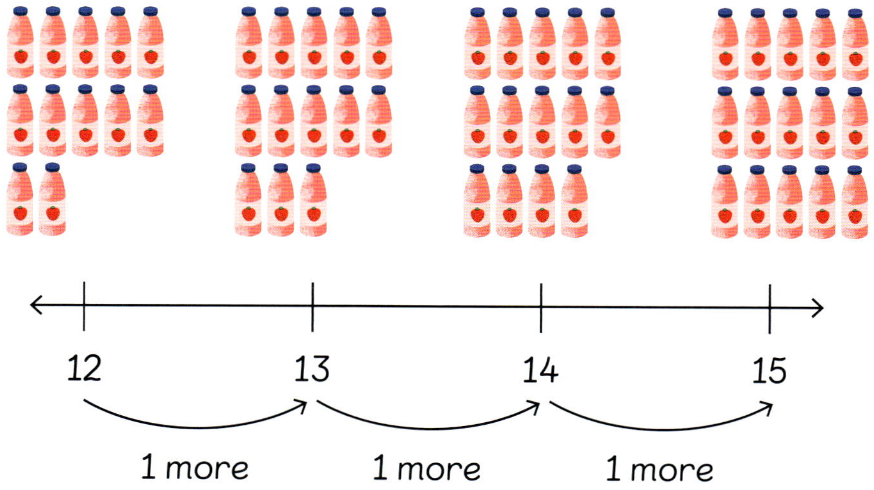

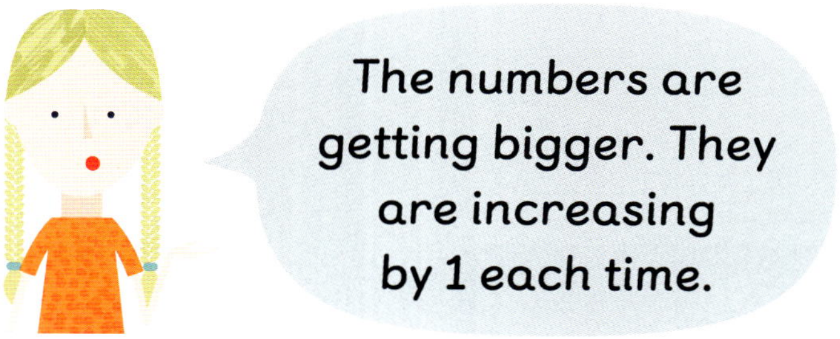

13 is 1 more than 12.
14 is 1 more than 13.
The next number is 15.
15 is 1 more than 14.

2

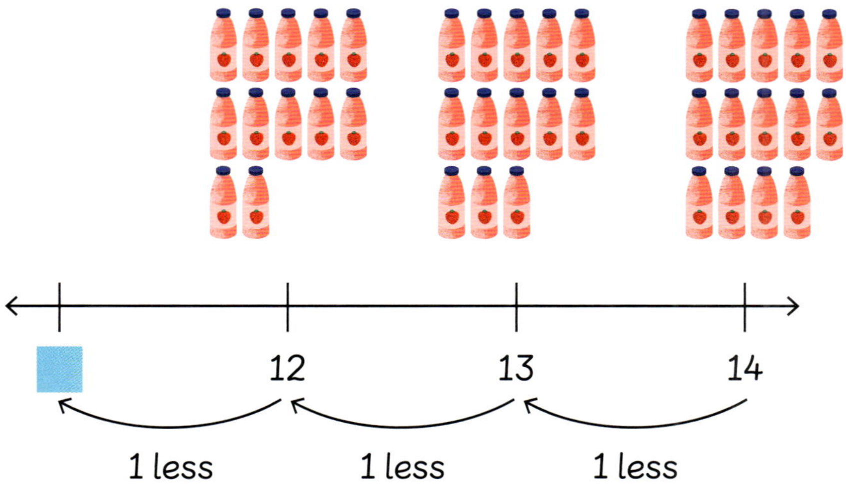

13 is 1 less than 14.
12 is 1 less than 13.
The next number is ▭ .

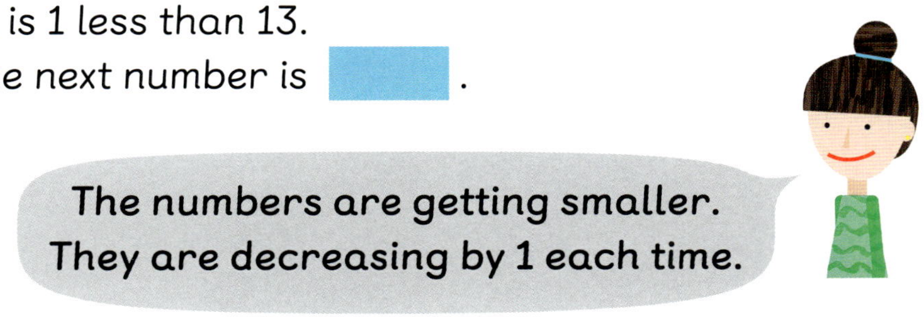

3 Describe the patterns the figures show.

(a)

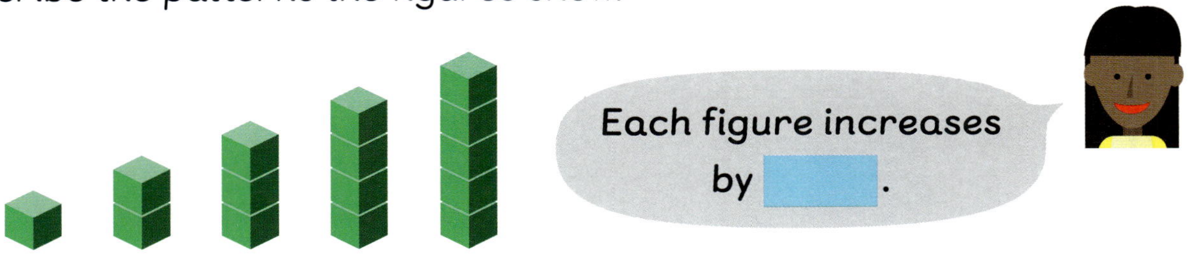

(b)

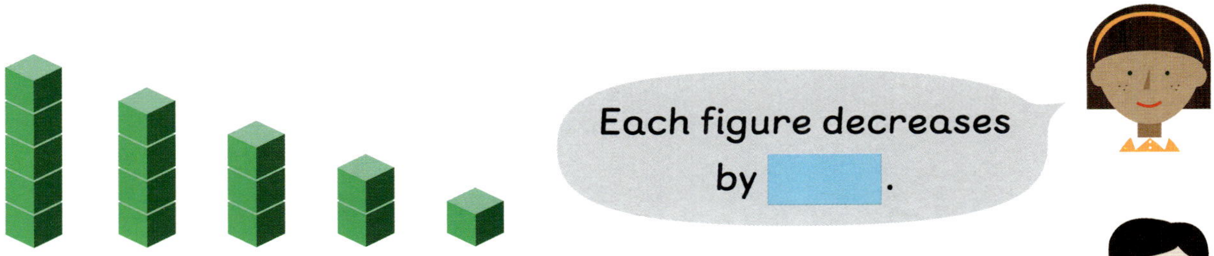

(c)

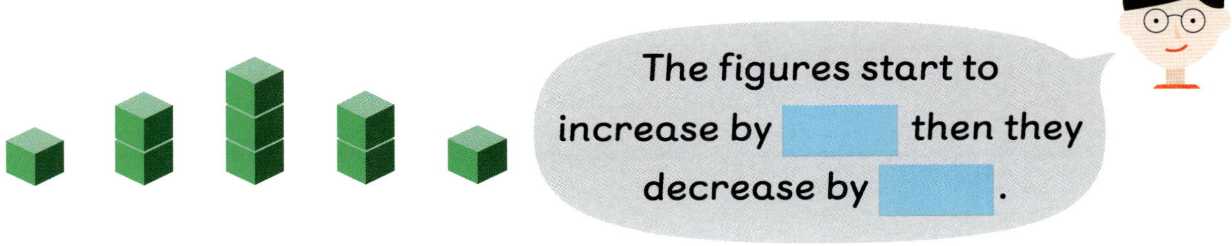

Activity Time

Partner work

You will need:

 40 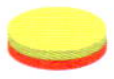 4 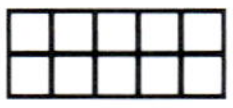

1. Pick a card and make the number using  and 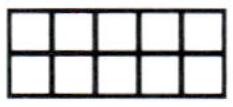.
2. Ask your partner to make the next number. They can choose to make the number 1 less or 1 more than yours.
3. Take turns.

Guided Practice

1 What are the missing numbers?

(a) 1 more than 8 is ____ .

(b) 1 less than 14 is ____ .

(c) ____ is 1 more than 16.

(d) ____ is 1 less than 10.

2 Help Ruby by completing the number patterns.

(a) 9, 10, 11, ____ , 13

(b) 12, 13, ____ , ____ , 16, 17

(c) 11, 10, 9, ____ , ____ , 6

(d) 9, ____ , 11, ____ , ____ , 14

Complete Worksheet **5**: pages **99–100**

Mind Challenge

Use the digit cards to complete the chart.

12 13 14 14
16 16 17

My Maths Journal

Ravi baked 8 cookies.
Ruby baked 15 cookies.
Sam baked 20 cookies.

Write two sentences to compare the number of cookies each child made.

For example: Ruby baked less cookies than Sam.

Review

I can...

- ☐ count up to 20.
- ☐ read numbers from 11 to 20.
- ☐ write numbers from 11 to 20.
- ☐ compare and order numbers up to 20.
- ☐ complete simple number patterns.

Chapter 7
Addition and Subtraction within 20

Add by Counting On

Lesson 1

Explore

How many sheep are there in total?

Master

1

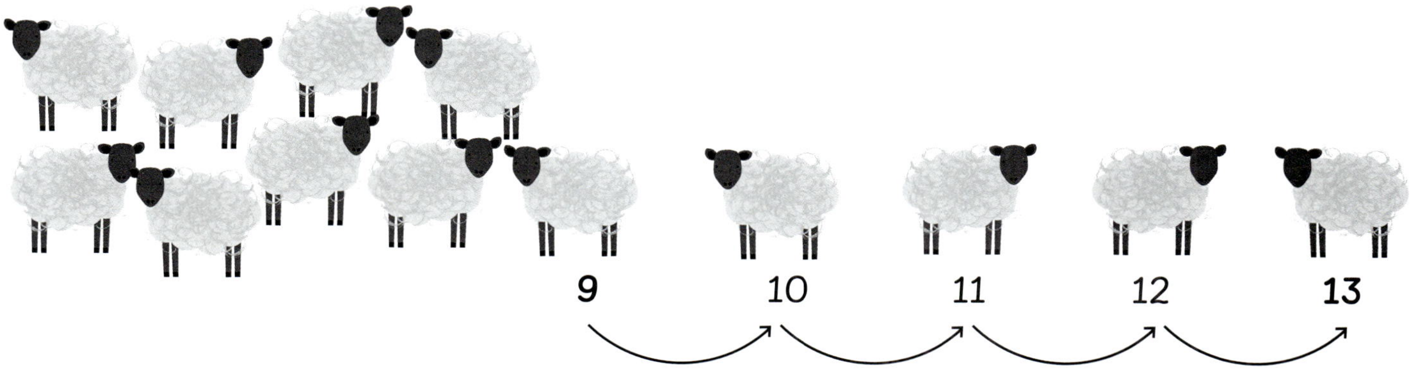

9	10	11	12	13

9 + 4 = 13

There are 13 sheep in total.

2 8 + 3 = ▭

8	9	10	11

8 + 3 = 11

Activity Time

Partner work

You will need:

2

1. Both partners place face down in a pile.
2. Pick two each from your piles.
3. Each add the two numbers together by counting on.
4. Check who has the greatest total.
5. Keep a record of your totals. Who has the greatest total most often?

Guided Practice

Use counting on to help you add.

1	2	3	4	5	6	7	8	9	10
11	12	13	14	15	16	17	18	19	20

1 8 + 5 =

2 13 + 6 =

3 4 + 12 =

4 7 + 11 =

Complete Worksheet **1**: pages **107–108**

Add by Making 10

Lesson 2

Explore

How many cows are in the field?

Master

1 8 + 4 =

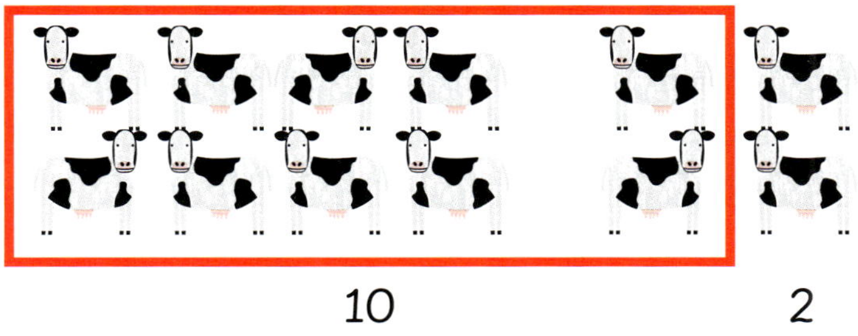

8 + 4 = 12

There are 12 cows in the field.

2 7 + 5 =

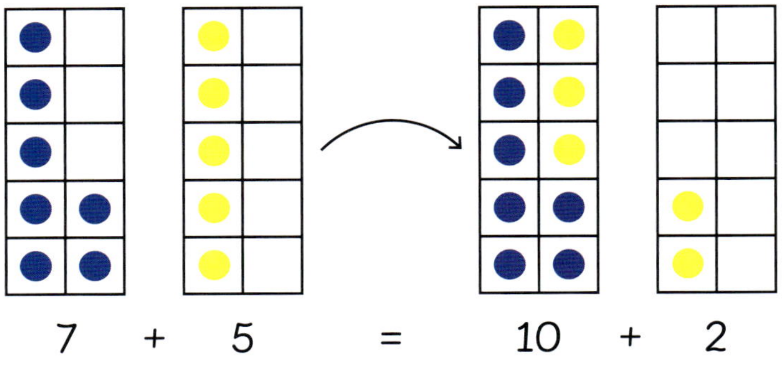

7 + 5 = 12

Activity Time

Partner work

You will need:
2 ten frames, 20 counters, number cards (5, 10)

1. Pick a card each.
2. Use counters to show your number on a ten frame.
3. Add your number and your partner's number together.

For example:

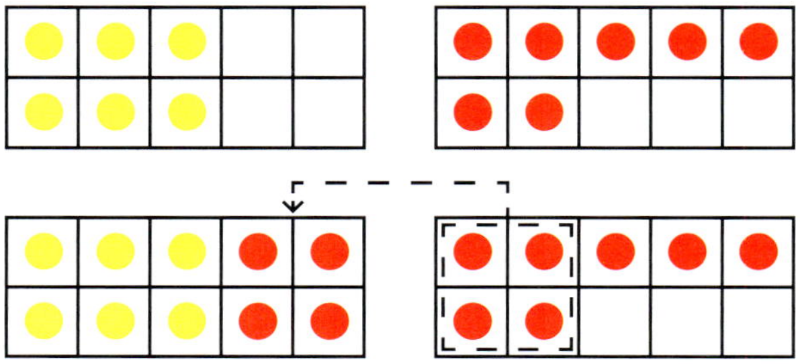

4. Repeat the activity.

Guided Practice

Make 10 and then add.

1

7 + 5 = ☐

2   5

5 + 7 = ☐

2   3

2

3 + 8 = ☐

8 + 3 = ☐

3

6 + 7 = ☐

7 + 6 = ☐

Complete Worksheet **2**: pages **109–112**

Add by Adding Ones

Lesson 3

Explore

How many hay bales can you see?

Master

11 + 5

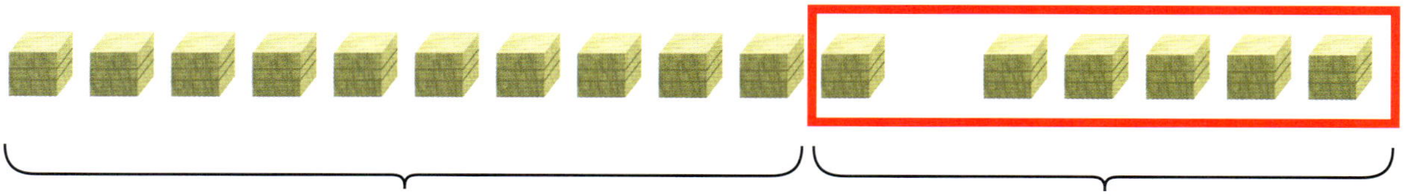

10 6

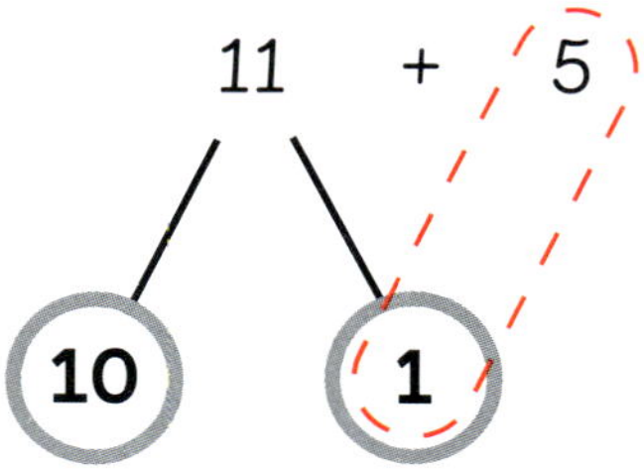

10 + 6 = 16

11 + 5 = 16

We can see 16 hay bales.

Guided Practice

Add.

1

12 + 5 = ____

12 → 10 and 2

5 + 12 = ____

12 → 2 and 10

2

16 + 3 = ____

16 → 10 and 6

3 + 16 = ____

16 → 6 and 10

3

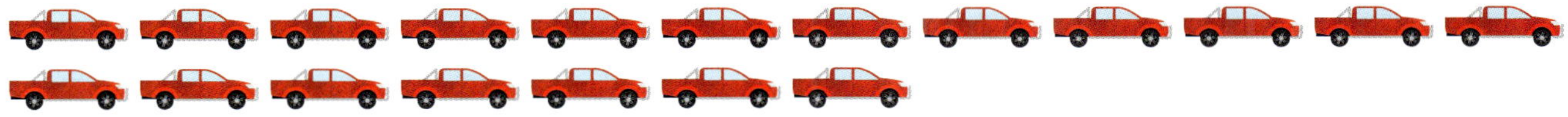

12 + 7 = ____

7 + 12 = ____

Complete Worksheet **3**: pages **113–114**

Subtract by Counting Back

Lesson 4

Explore

Master

16 − 4 = ▭

There were 16 pumpkins altogether.
Holly and Charles took 4 pumpkins.

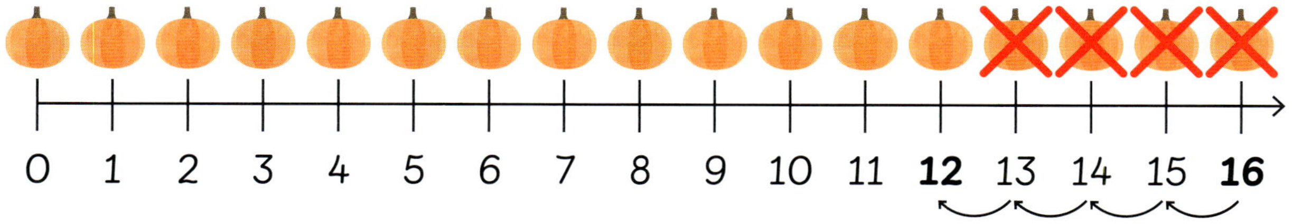

16 − 4 = 12

12 pumpkins were left.

Activity Time

Partner work

You will need:

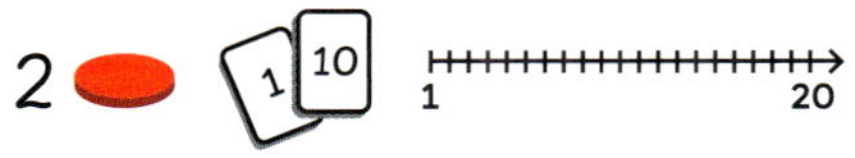

1. Place your on any number from 11 to 20.
2. Your partner picks a 1 10.
3. Count back the number of steps on the card.
4. Take turns.

Guided Practice

Subtract the ones by counting back.

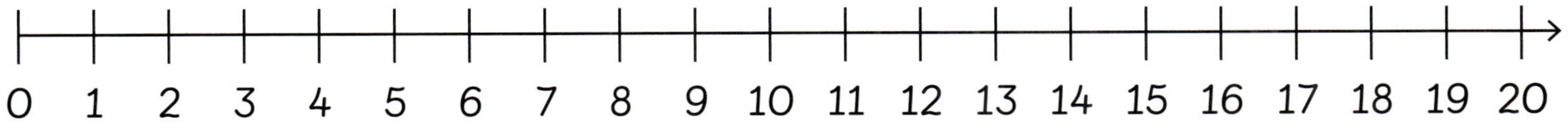

1 14 − 3 = ☐

2 16 − 4 = ☐

3 13 − 5 = ☐

4 12 − 7 = ☐

Complete Worksheet 4: pages **115–116**

Subtract by Subtracting Ones

Lesson 5

Explore

Four hens walk into their house.
How many hens are left outside?

Master

1 $15 - 4 = 11$

First | Then | Now

15 | -4 | 11

First there were 15 hens.
Then 4 went into the house.
Now there are 11 hens.
$15 - 4 = 11$

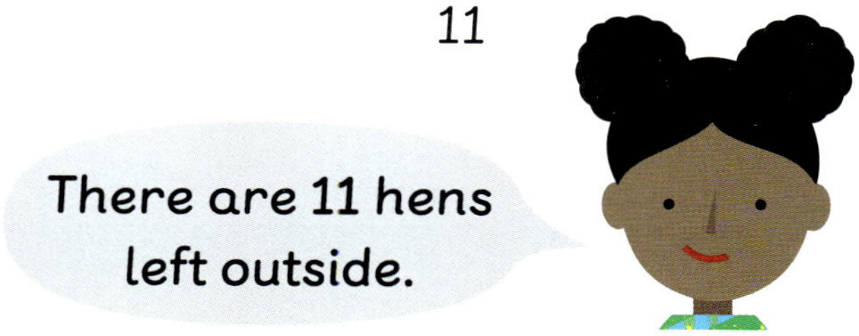

2 15 − 4 =

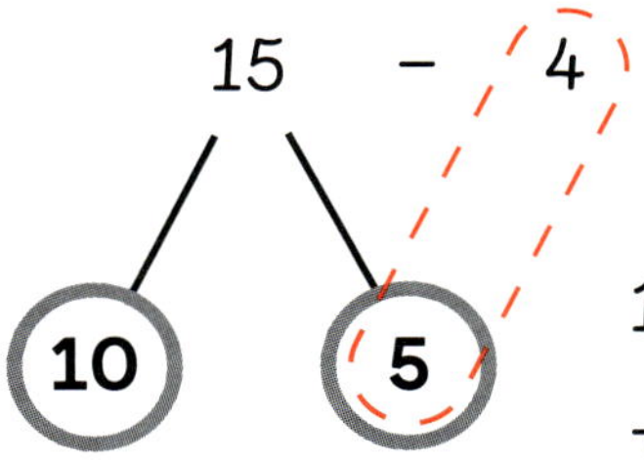

15 − 4 = 11

There are 11 hens left outside.

3 17 − 4 =

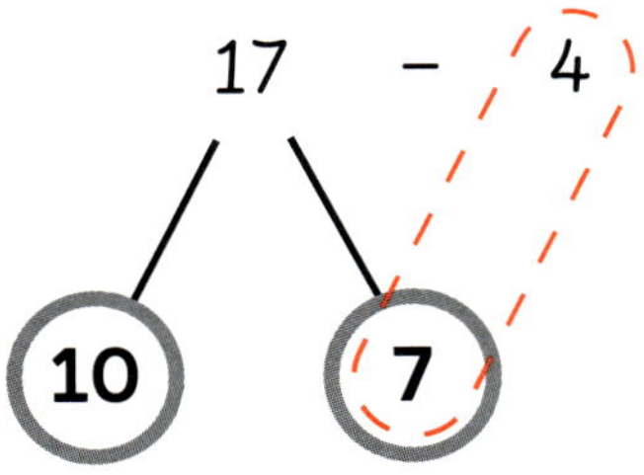

17 − 4 = 13

There are 13 eggs left.

Activity Time

Partner work

You will need:

① Pick one  and one .

② Work together to subtract the smaller number from the greater number.

③ Ask your partner to write the equation.

④ Take turns.

Guided Practice

Subtract.

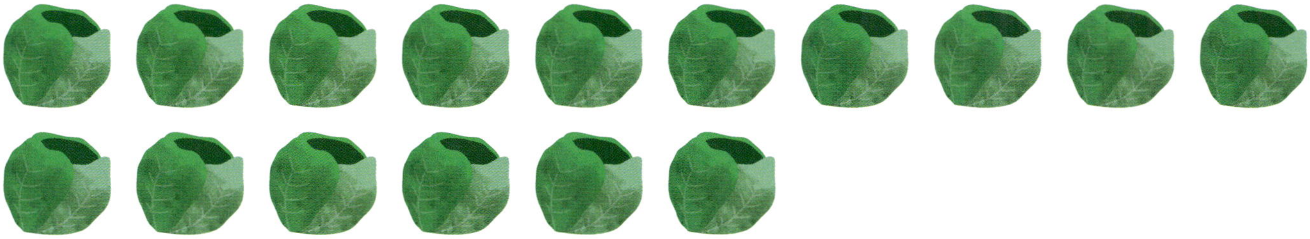

1 16 − 4 =

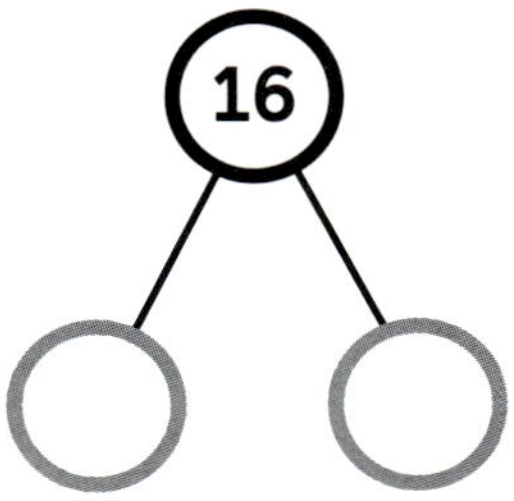

2 16 − 6 =

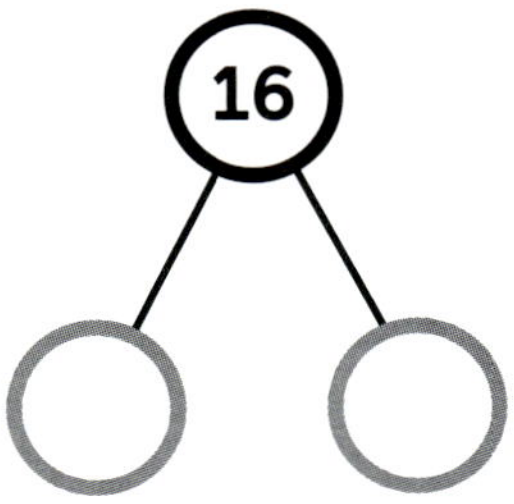

3 16 − 3 =

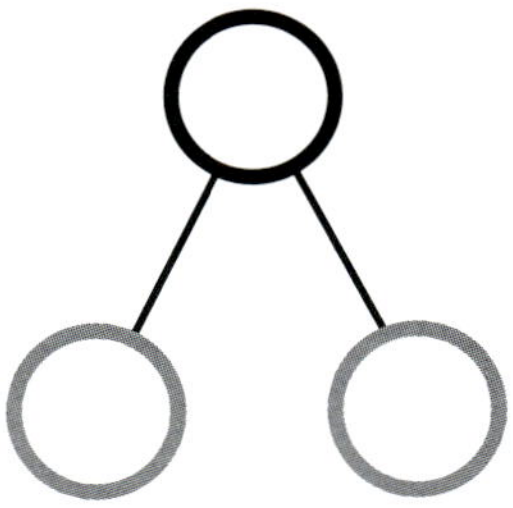

Complete Worksheet **5**: pages **117–118**

Subtract from 10

Lesson 6

Explore

There are 16 logs altogether.
9 logs are taken away.
How many logs are left?

Master

16 − 9 = ▭

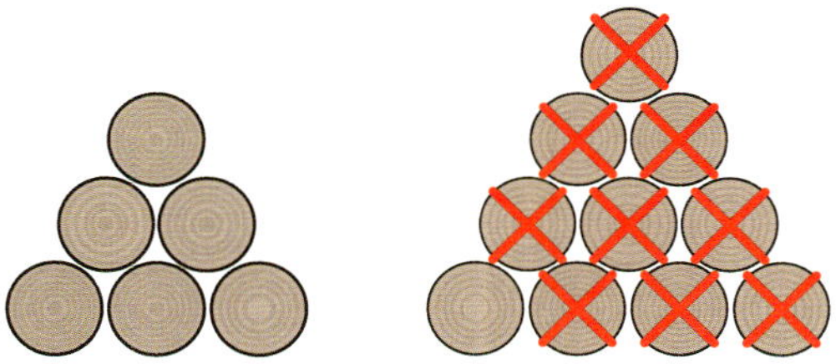

6 10

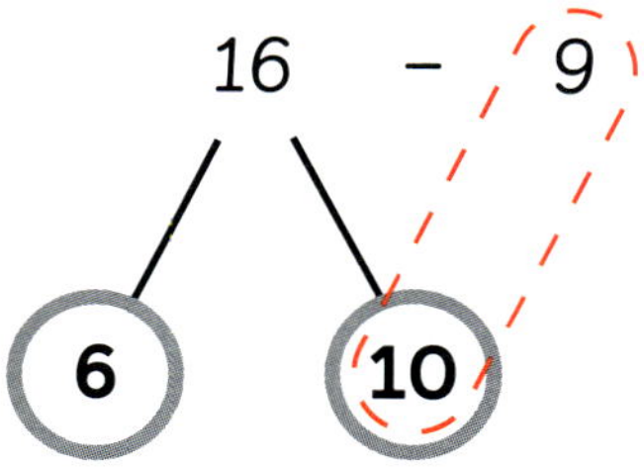

10 − 9 = 1
1 + 6 = 7

16 − 9 = 7

There are 7 logs left.

Guided Practice

Subtract.

1 (a) 14 − 6 = ___

10 − 6 = ___ ___ + 4 = ___

(b) 13 − 7 = ___

10 − 7 = ___ ___ + 3 = ___

(c) 11 − 5 = ___

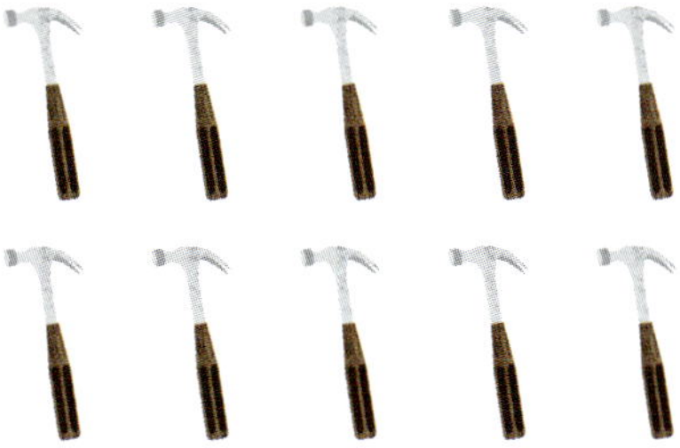

10 − 5 = ___ ___ + ___ = ___

2 (a) 12 − 4 = ___ (b) 15 − 8 = ___

Complete Worksheet **6**: pages **119–120**

Addition and Subtraction Facts

Lesson 7

Explore

Is Ravi correct?

Master

1

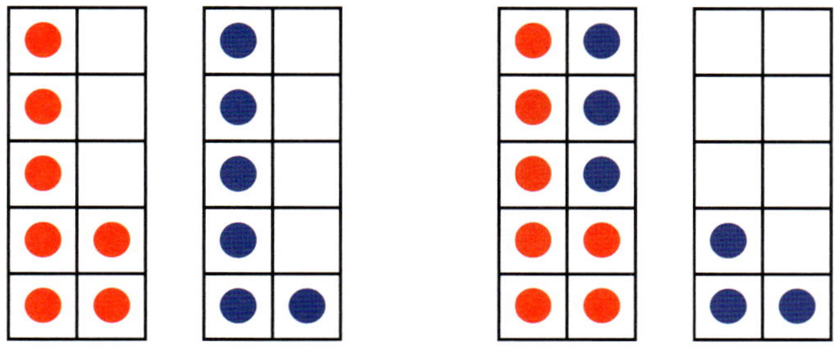

7 + 6 = 10 + 3
= 13

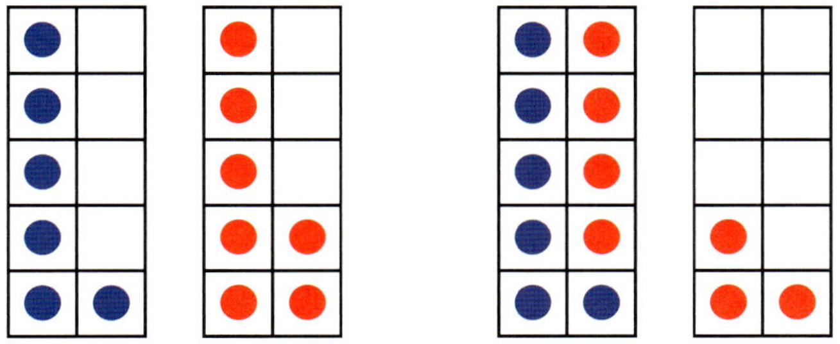

6 + 7 = 10 + 3
= 13

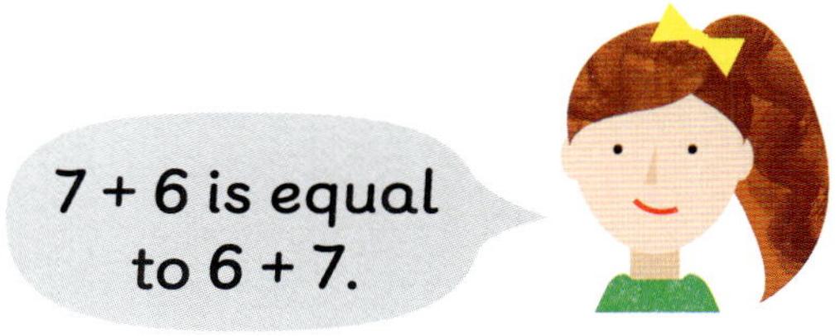

2

7 + 6 = 13

6 + 7 = 13

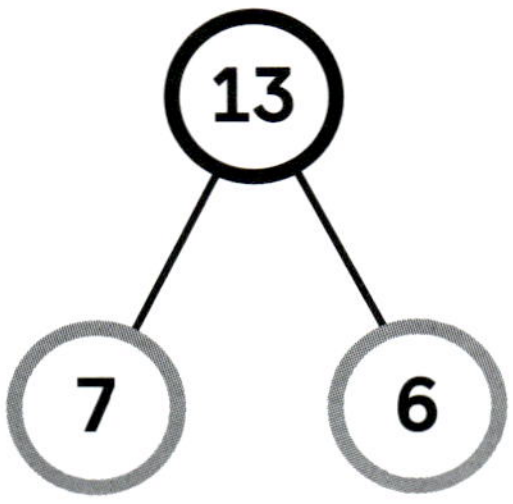

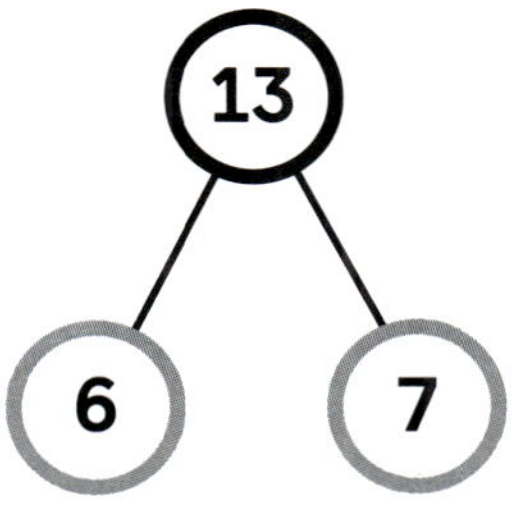

13 − 6 = 7

13 − 7 = 6

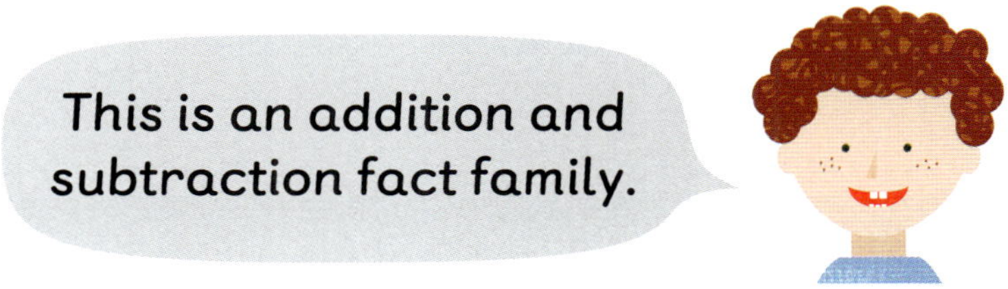

Guided Practice

1 Complete the addition and subtraction fact families.

(a) 4 + ___ = 5   ___ − 1 = 4

1 + ___ = 5   5 − ___ = 1

(b) 2 + ___ = 5   5 − ___ = ___

3 + ___ = 5   5 − ___ = ___

2 Complete the addition and subtraction fact families.

(a) 5 + 8 = 13

8 + 5 = ___

13 − 5 = ___

13 − 8 = ___

(b) 6 + 9 = ___

9 + 6 = ___

15 − 9 = ___

15 − 6 = ___

3 Use the numbers to make an addition and subtraction fact family.

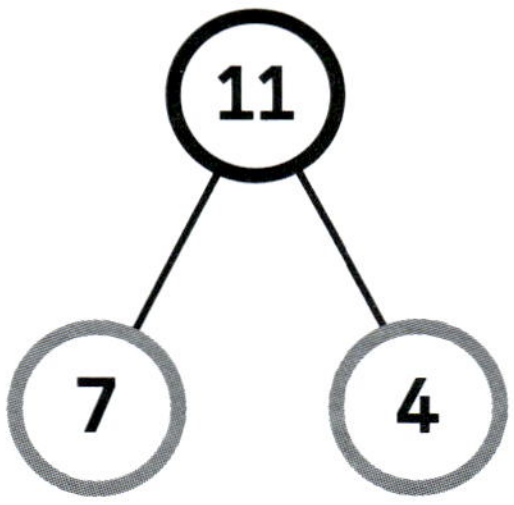

___ + ___ = ___   ___ − ___ = ___

___ + ___ = ___   ___ − ___ = ___

Complete Worksheet **7**: pages **121–122**

Mind Challenge

What numbers are the hands covering?

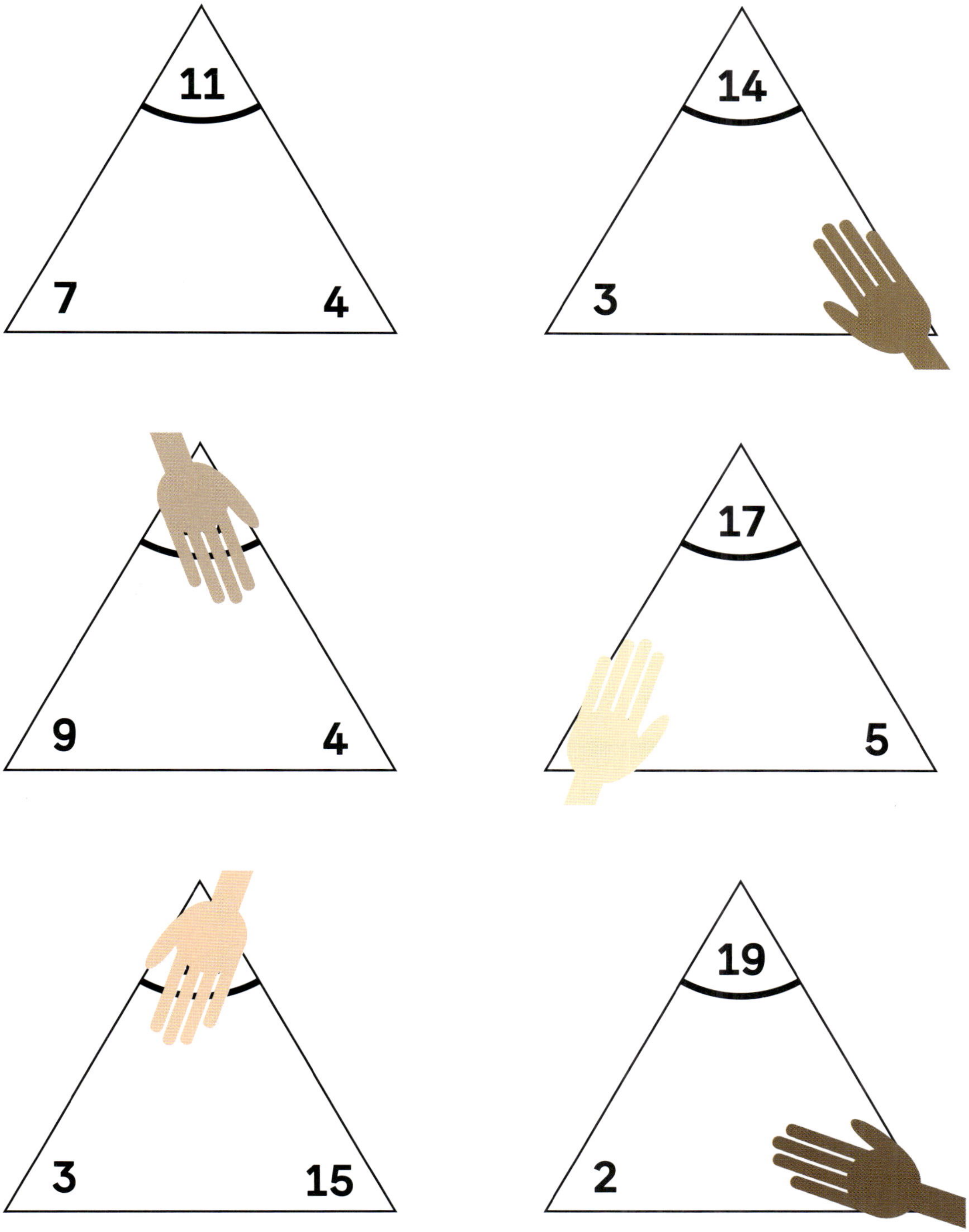

My Maths Journal

Explain to Amira how you can use 14 − 5 = 9 to solve 9 + 5 = 14.
Use pictures to help.

Review

I can...

- ☐ count on to add.
- ☐ add by making 10.
- ☐ add by adding ones.
- ☐ count back to subtract.
- ☐ subtract by subtracting ones.
- ☐ subtract by subtracting from 10.
- ☐ make addition and subtraction fact families.

Can you name the 2D shapes in the playground?

Chapter 8
Shapes and Patterns

Recognising 3D Shapes

Lesson 1

Explore

Can you name the 3D shapes you see at the beach?

Master

1 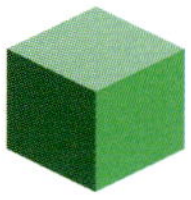 is a cube.

These are also cubes.

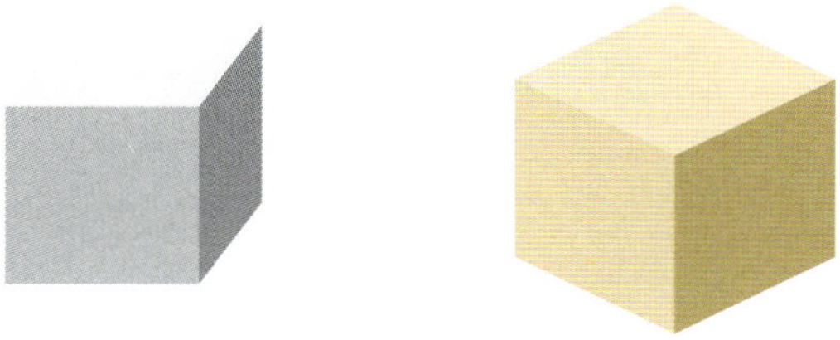

2 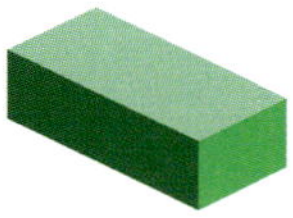 is a cuboid.

These are also cuboids.

3  is a sphere.

These are also spheres.

4  is a pyramid.

These are also pyramids.

Activity Time

Partner work

You will need:

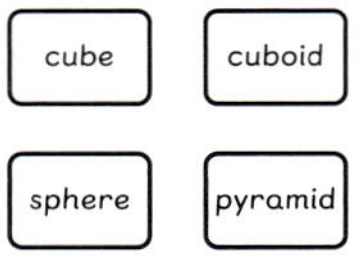

You are 3D shape detectors.

1. Look around your classroom for 3D shapes: cubes, cuboids, spheres and pyramids.
2. Put the correct label beside the object.
3. How many 3D shapes can you find?

Guided Practice

Match each name to a 3D shape.

sphere

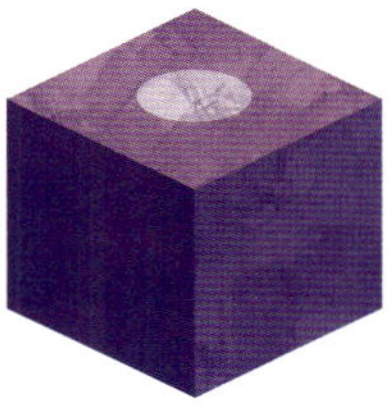

cube

cuboid

pyramid

Complete Worksheet **1**: pages **125–126**

Recognising 2D Shapes

Lesson 2

Explore

There is more than one rectangle in this playground.

Is Emma correct?
What other 2D shapes can you find?

Master

1

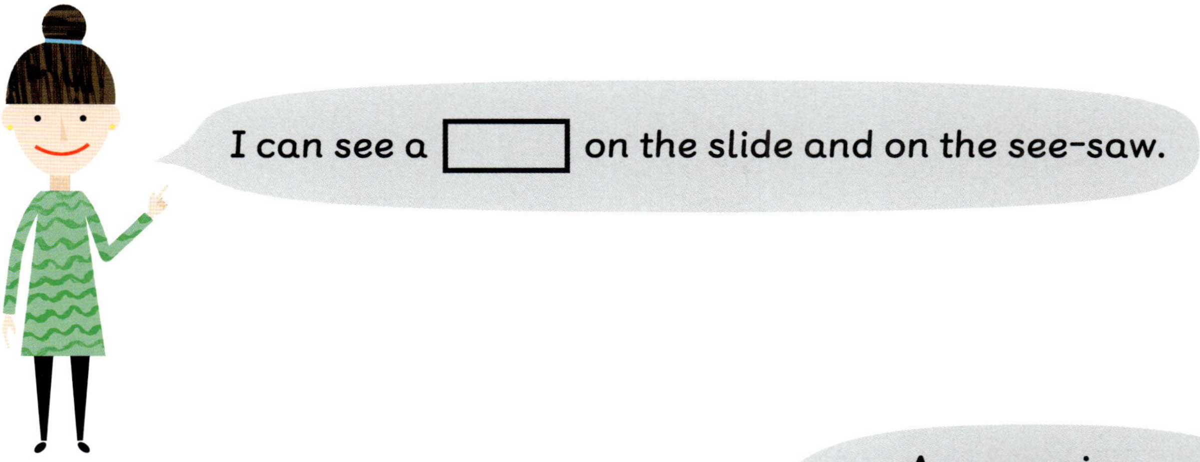

These 2D shapes are also rectangles.

2 The frame of the swing is a $\triangle$.

These 2D shapes are also triangles.

3 The climbing frame has a .

These 2D shapes are also squares.

4 The merry-go-round has a big 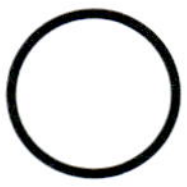 as a base.

These 2D shapes are also circles.

Activity Time

Partner work

You will need:

square | rectangle | circle | triangle

You are 2D shape detectors.

1. Look around your classroom for 2D shapes: squares, rectangles, circles and triangles.
2. Put the correct label beside the object.
3. Which object shows more than one 2D shape?
4. Which 2D shape was difficult to find?

Guided Practice

1 How many circles can you find?

2 How many squares can you find?

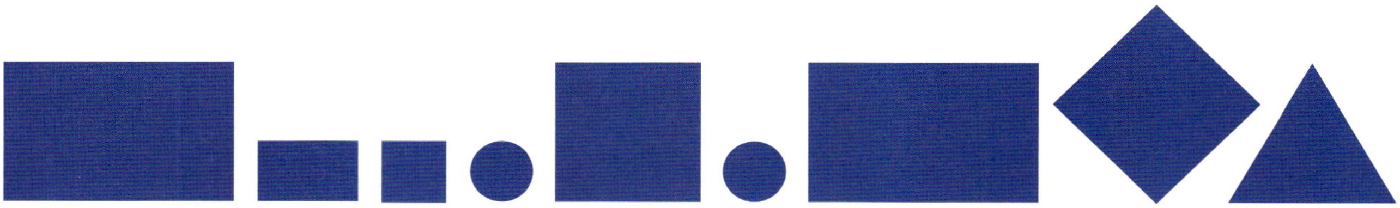

3 How many rectangles can you find?

4 How many triangles can you find?

Complete Worksheet **2**: pages **127–128**

Grouping 2D Shapes

Lesson 3

Explore

How can  and  group these 2D shapes?

Elliott Ruby

Master

1

I can group them by number of sides.

2

I can group them by the length of their sides.

3

I can group them by number of corners.

Activity Time

Partner work

You will need:

1. Use some  to make a group.
2. Ask your partner how your  are grouped.
3. Take turns.

Guided Practice

Are the statements **true** or **false**?

1

The 2D shapes are grouped by **number of corners**.

2

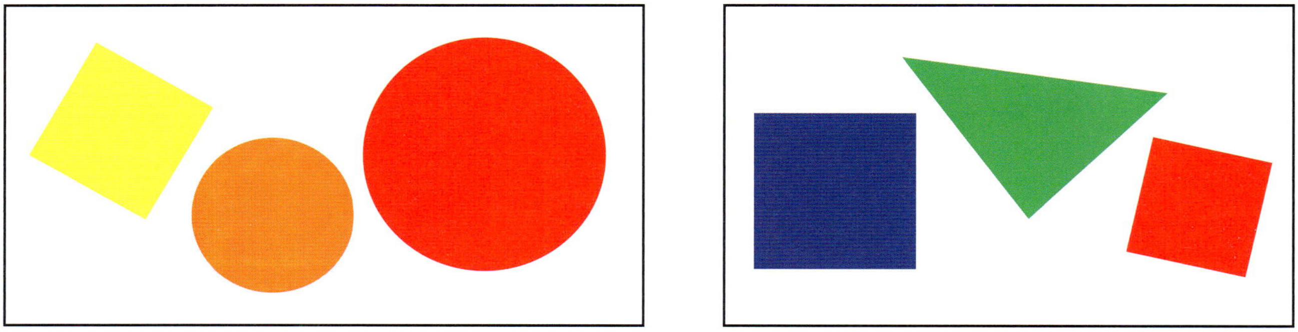

The 2D shapes are grouped by **shape**.

3

The 2D shapes are grouped by **number of sides**.

4

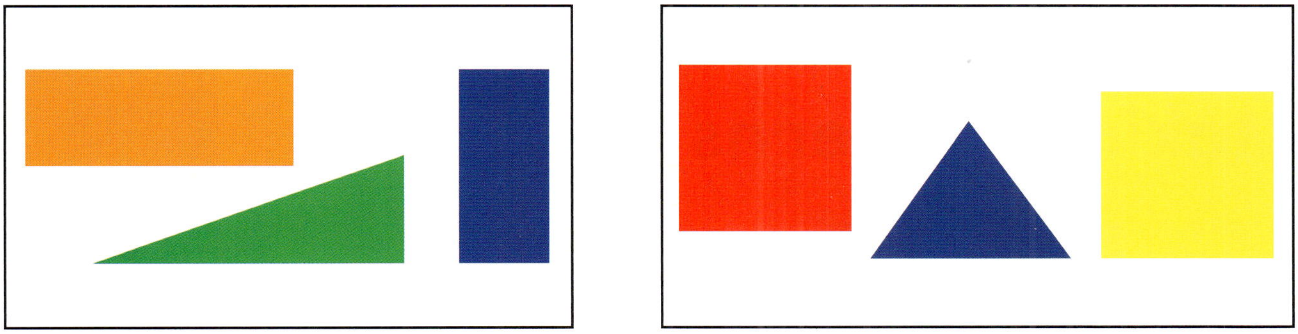

The 2D shapes are grouped by **the length of their sides**.

5

The 2D shapes are grouped by **shape**.

6

The 2D shapes are grouped by **shape**.

Complete Worksheet **3**: pages **129–130**

Making Patterns

Lesson 4

Explore

Holly's dad put wallpaper on one wall in her room.

How can we describe the patterns on the wallpaper?

Master

1

 green circle, yellow triangle	,  green circle, yellow triangle ...

2

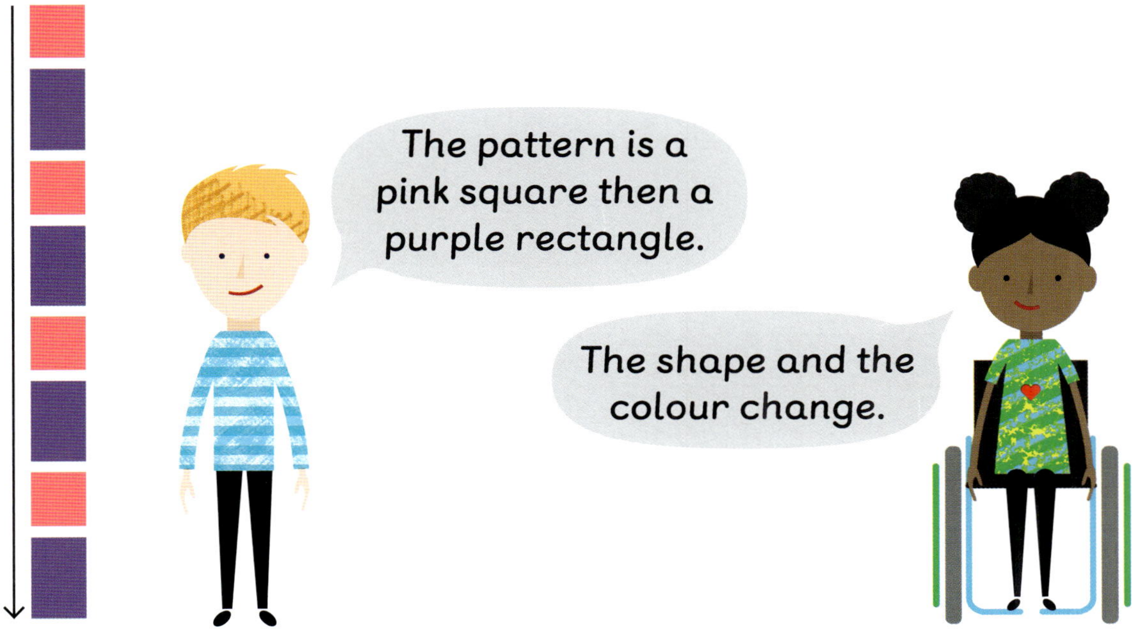

 pink square, purple rectangle	,  pink square, purple rectangle ...

3 Describe the pattern on this wallpaper.

big orange triangle,
small green triangle

,

big orange triangle,
small green triangle

...

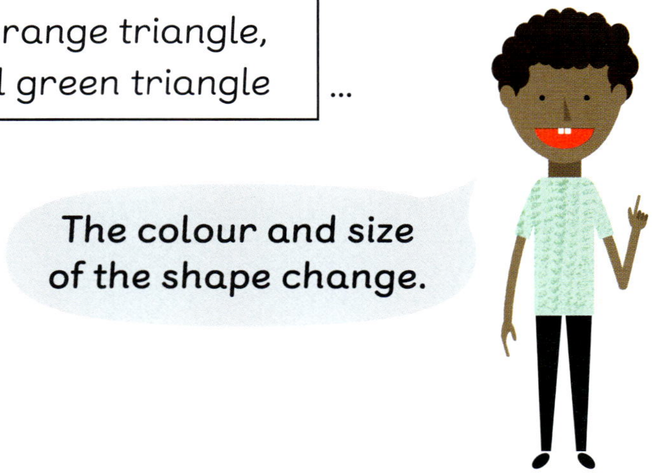

Guided Practice

1 Complete.

(a)

(b)

(c)

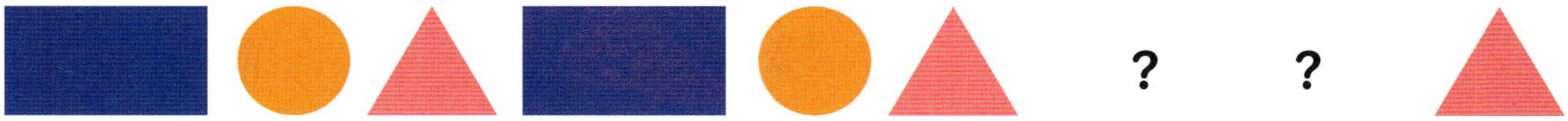

2 Use the following shapes to make your own pattern.

(a)

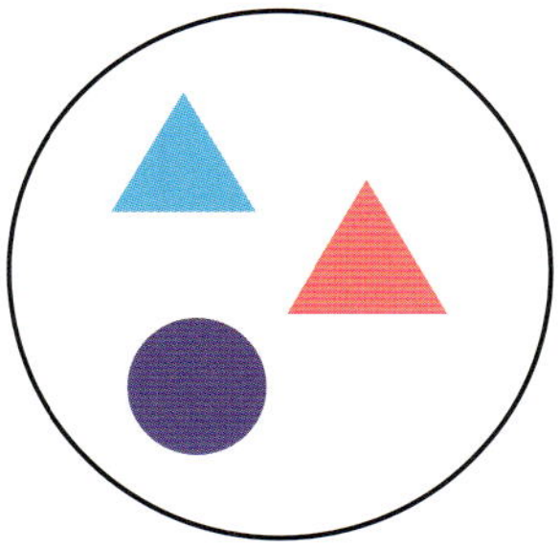

____

(b)

____

Complete Worksheet **4**: pages **131–132**

Mind Challenge

Ruby designed a picture frame using patterns.

How would you use patterns on your picture frame?

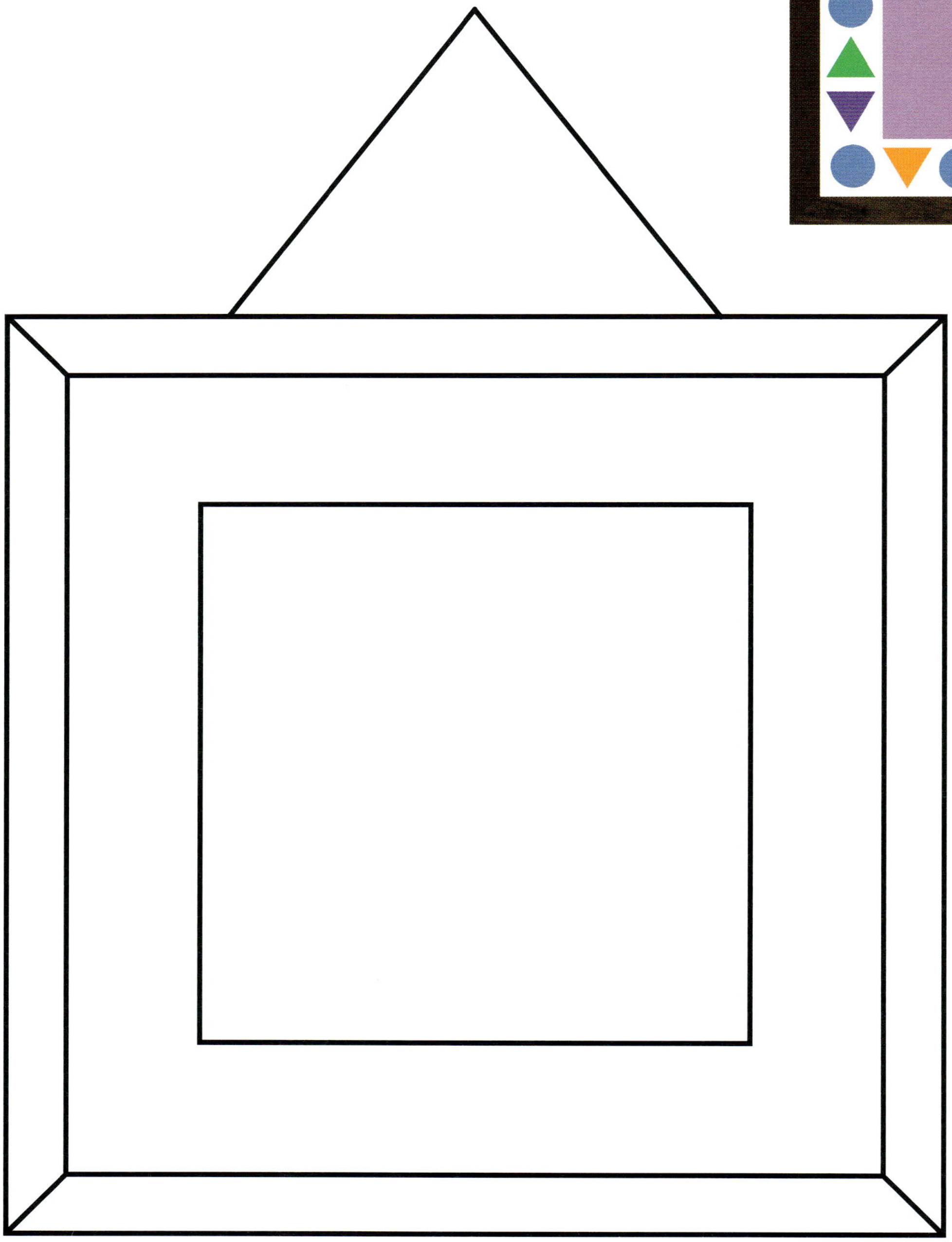

My Maths Journal

Think of patterns that use shapes in everyday life.
Draw the pattern and write where you have seen it.

Review

I can...

- ☐ find and name 3D shapes and 2D shapes.
- ☐ find 2D shapes in 3D shapes.
- ☐ group 2D shapes.
- ☐ make and complete patterns with 2D shapes.

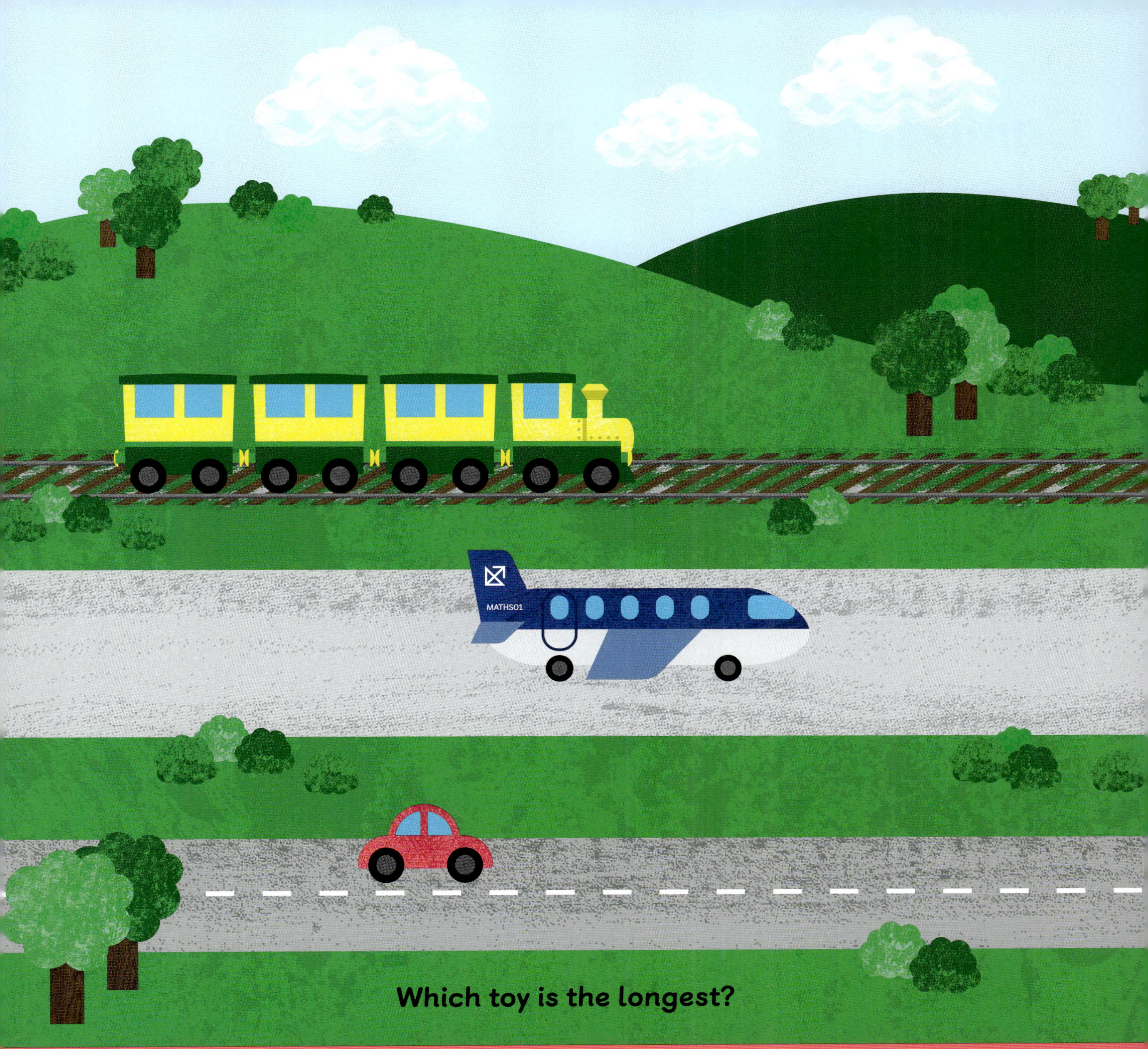

Which toy is the longest?

Chapter 9

Height and Length

Comparing Height and Length

Lesson 1

Explore

Compare the heights of the animals.

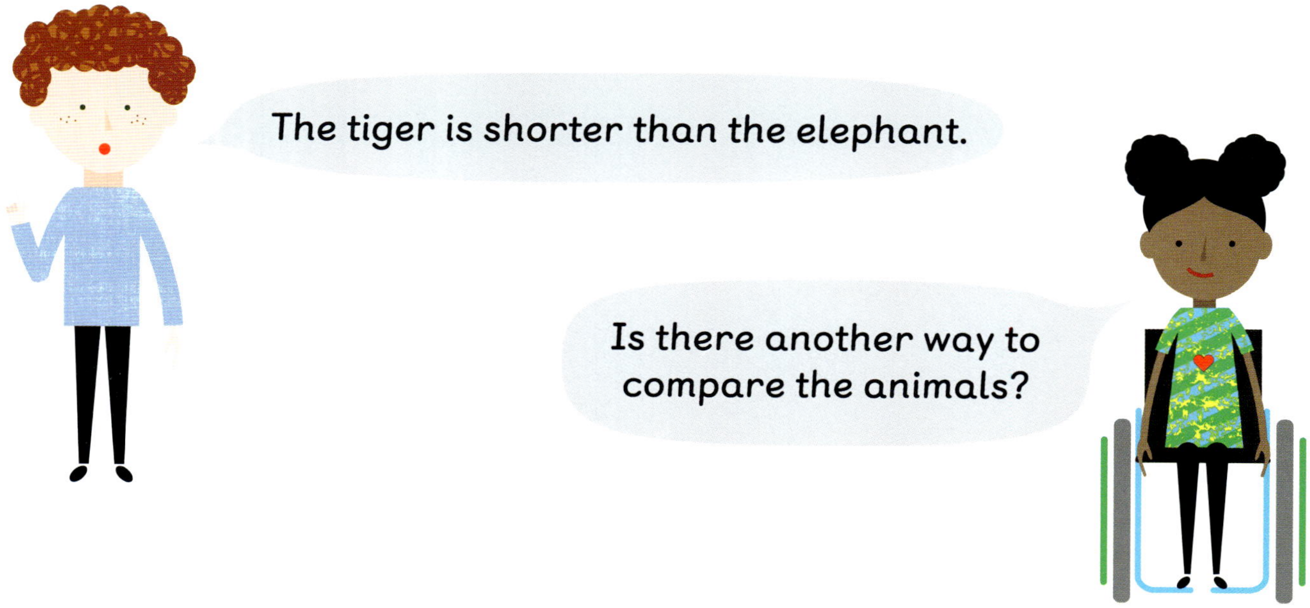

Master

1

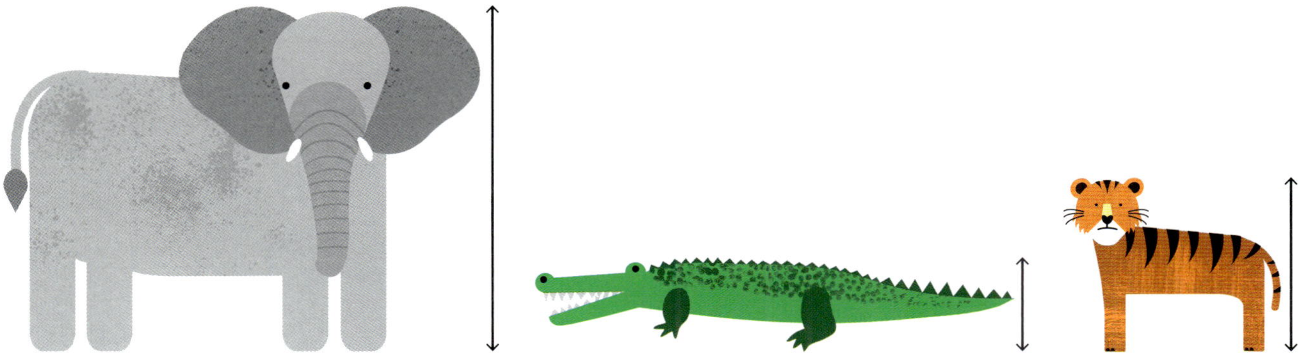

The  is shorter than the  .

The  is shorter than the  .

The  is the shortest.

The  is taller than the  .

The  is taller than the  .

The  is the tallest.

shortest ⟷ tallest

2 Compare the lengths of the toys.

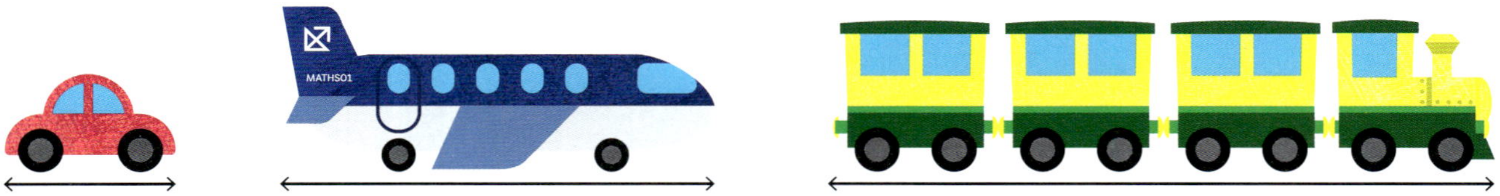

The  is longer than the  .

The  is longer than the  .

The  is the longest.

The  is shorter than the  .

The  is shorter than the  .

The  is the shortest.

Guided Practice

Use the words **longer**, **shorter** or **taller** to complete the sentences.

1 The  is ________ than the  .

2 The  is ________ than the  .

3 This  is ______ than this .

4 The  is ______ than the  and ______ than the .

Complete Worksheet **1**: pages **137–140**

Measuring Length Using Things

Lesson 2

Explore

How can  use rubbers to measure the length of the bookmark?

Master

1 We can use ____ to measure things.

The length of the  is about 3 ____ long.

The length of the 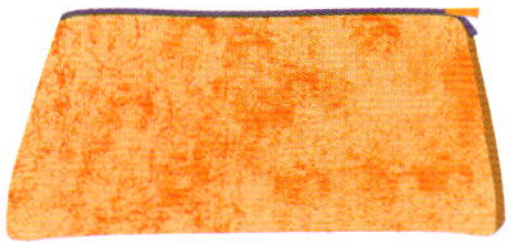 is about 4 long.

The 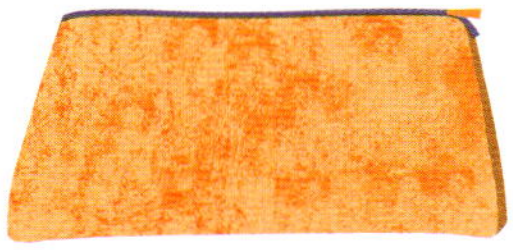 is longer than the .

The  is shorter than the 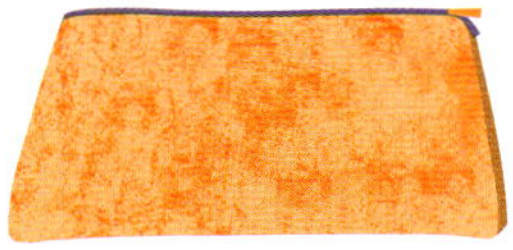.

2

The length of the  is about 4 ▢ long.

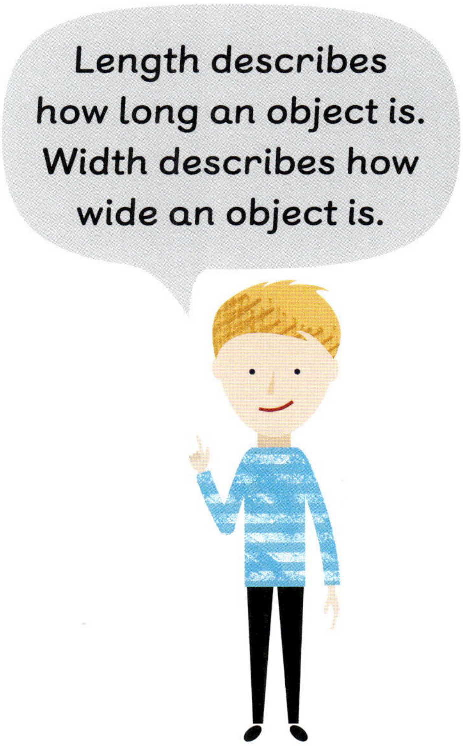

The width of the  is about 5 ▢ wide.

Activity Time

Partner work

Use to measure:

(a) scissors

(b) a pencil

(c) a pencil case

You will need:

Guided Practice

1 What is the length of the dragonfly?

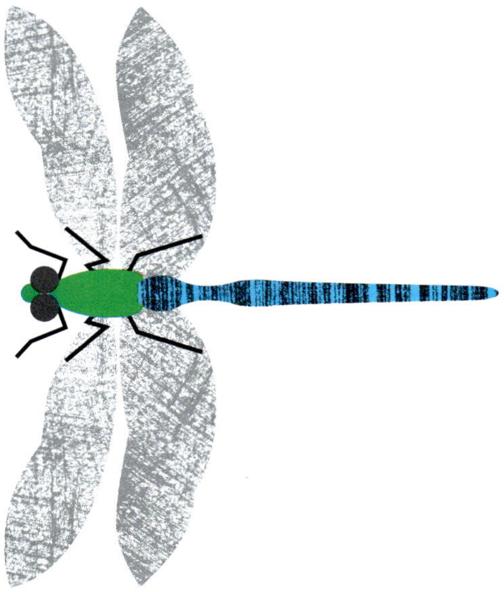

The 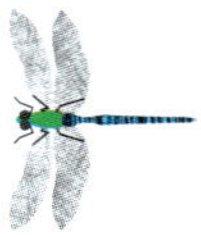 is about ______ long.

2 What is the length of the grasshopper?

The  is about ______ ______ long.

3 What is the length of the paintbrush?

The  is about ______ ______ long.

Complete Worksheet **2**: pages **141–144**

Measuring Height and Length Using Body Parts

Explore

Can we use our  or 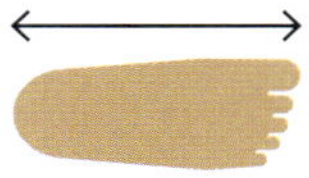 to measure the length of this table?

Master

1

We can use our hands and feet to measure.

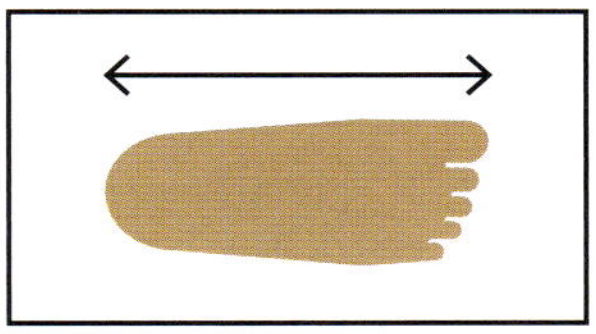

The table is about 6  long.

2

The pot plant is about 10  tall.

Activity Time

Partner work

Take turns to use your  or 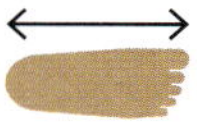 to measure:

(a) the width of your maths book

(b) the height of your chair

(c) the length of your pencil case

Record your measurements using a table.

You will need:

your 

your 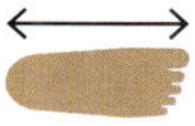

Object	Number of	
		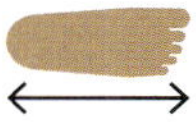

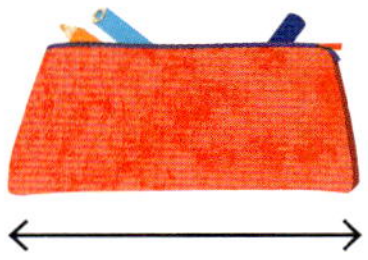		

Guided Practice

Measure the objects.

1 (a) What is the height of the wheat?

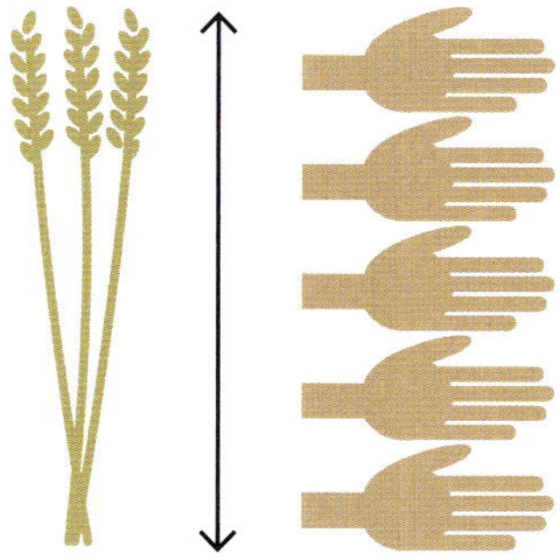

The wheat is about ______ 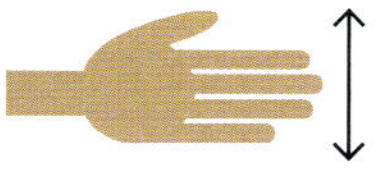 tall.

(b) What is the length of the window?

The window is about ______ 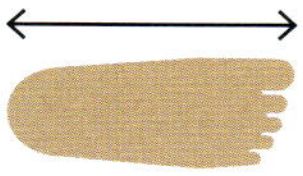 long.

2 (a)

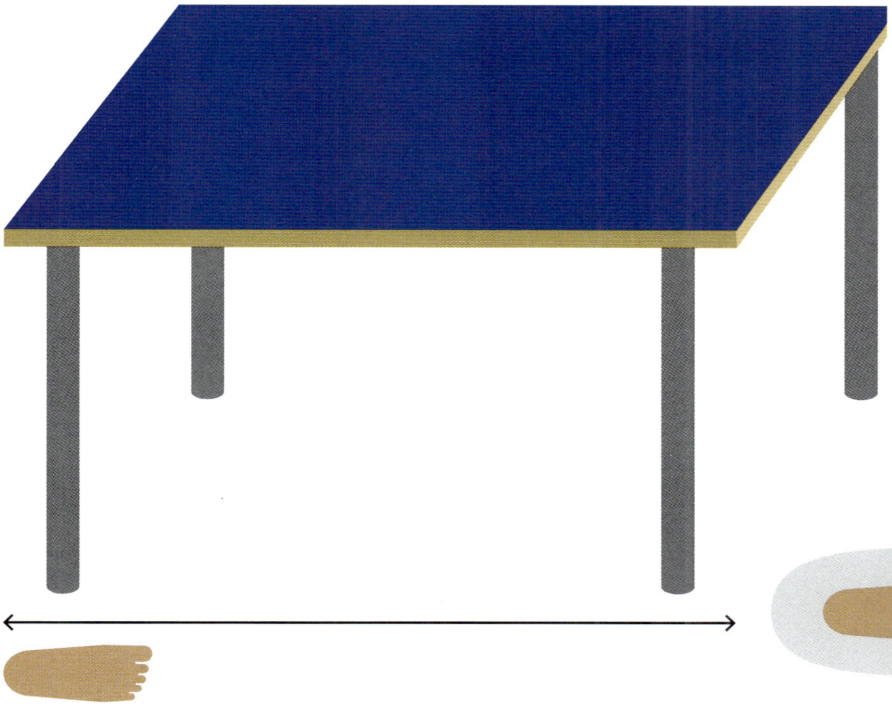

The desk is about ___  long.

The length of the desk is about ___ units.

(b)

The bookshelf is about ___  tall.

The height of the bookshelf is about ___ units.

Complete Worksheet 3: pages 145–148

Measuring Height and Length Using a Ruler

Lesson 4

Explore

Ravi is using tiles to measure the length of the sticker. How could he find the width of the bookmark?

Master

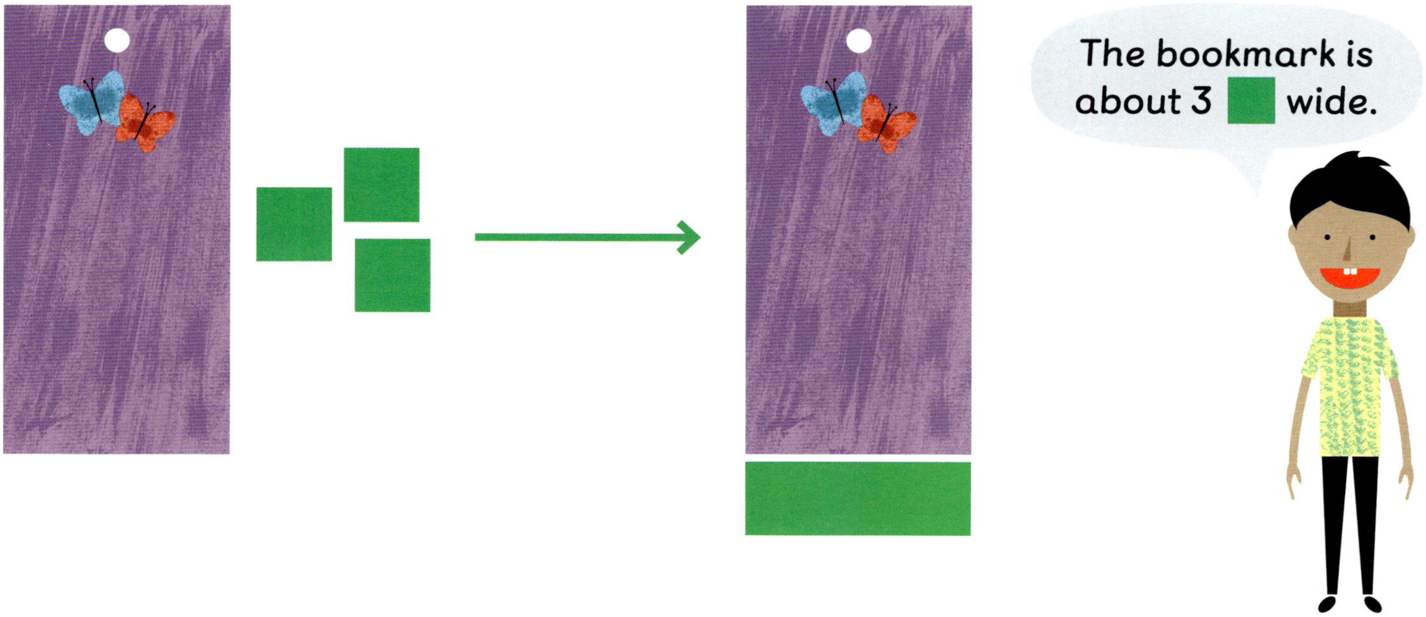

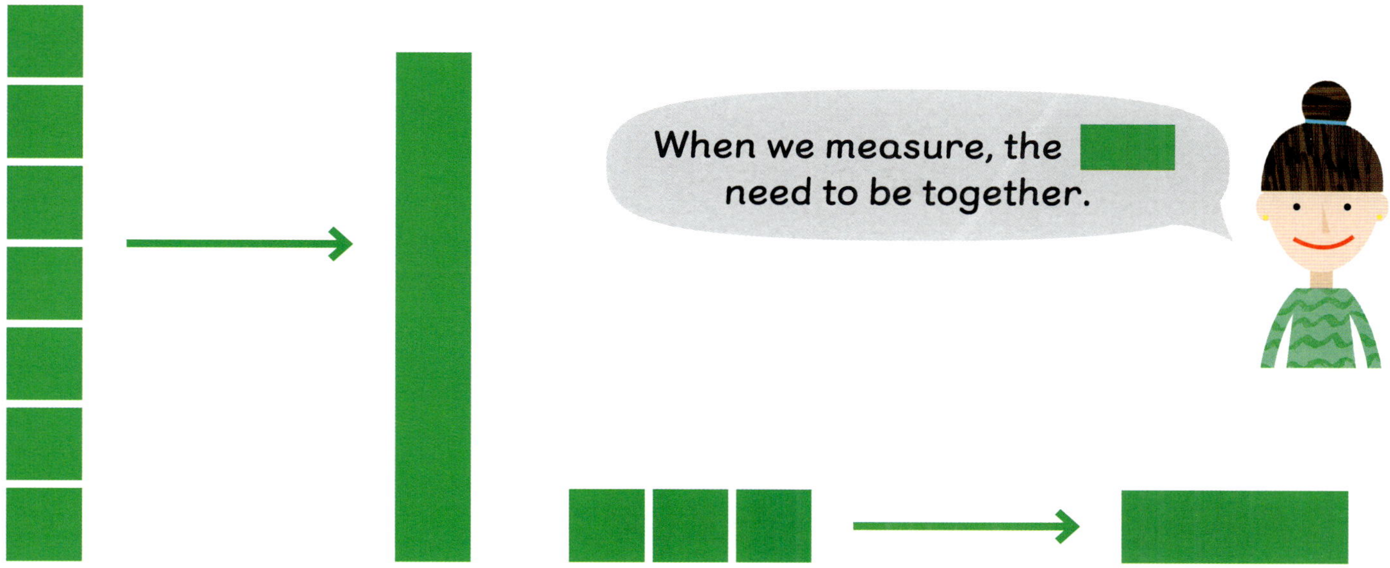
When we measure, the
need to be together.

1 cm
1 cm
This is 1 centimetre tall
and 1 centimetre wide.

The sticker is about
7 centimetres long.
The bookmark is about
3 centimetres wide.

Activity Time

Partner work

 and  made a ruler to measure the length and height of different items in their classroom.

You will need:

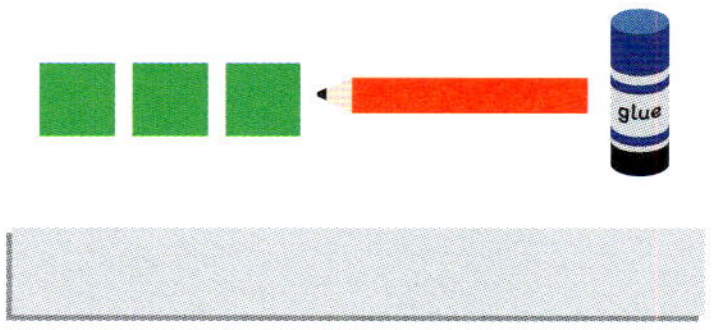

1. Get your 1-cm tiles and strip of paper.

2. Make a ruler.

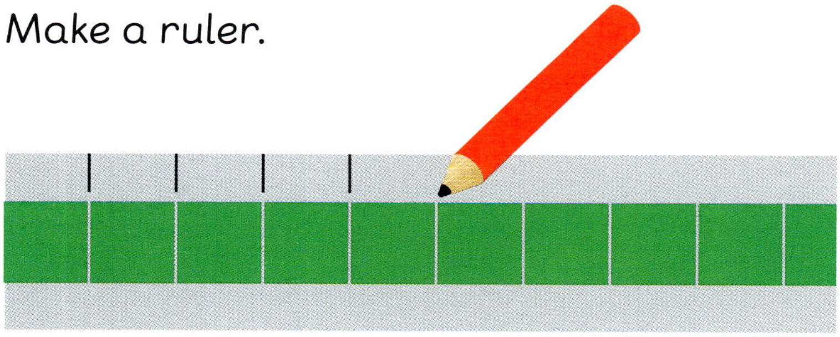

3. Use your ruler to measure items in your classroom.

The pencil sharpener is about 3 centimetres.

4. Use your ruler for **Guided Practice**.

Guided Practice

1 Use the ruler to measure these lines.

(a)

The line is about ☐ centimetres long.

(b)

The line is about ☐ centimetres long.

2

 is about ☐ centimetres long.

3

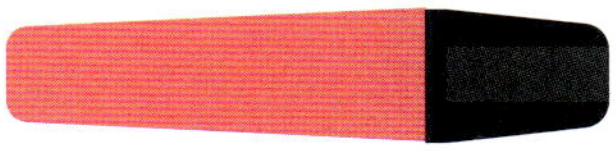 is about ☐ centimetres long.

4

 is about ☐ centimetres long.

Complete Worksheet 4: pages **149–152**

Mind Challenge

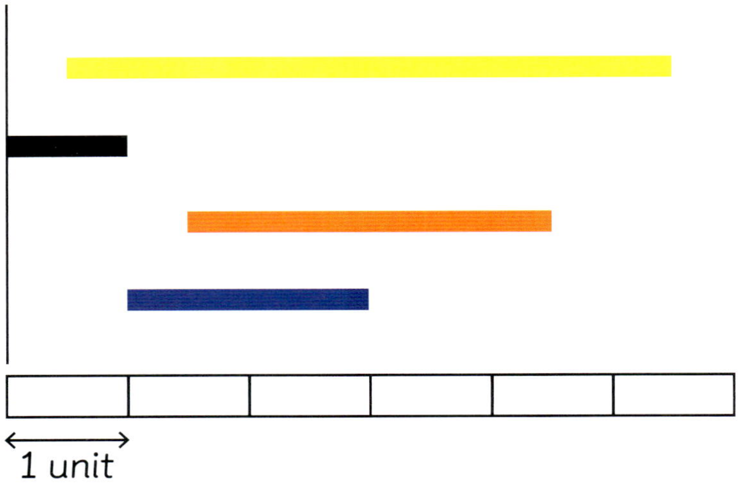

Write down the lengths of the lines and complete the sentences.

The yellow line is the ________.

The orange line is the same length as the ________ line and the ________ line put together.

My Maths Journal

Which caterpillar is longer?
Explain your answer.

Review

I can...

- ☐ compare the height and length of objects.
- ☐ measure the height and length of objects.